Mobile Opportunistic Networks: Architectures, Frameworks and Applications

Mobile Opportunistic Networks: Architectures, Frameworks and Applications

Edited by **Grant Harper**

C WILLFORD PRESS

New York

Published by Willford Press,
118-35 Queens Blvd., Suite 400,
Forest Hills, NY 11375, USA
www.willfordpress.com

Mobile Opportunistic Networks: Architectures, Frameworks and Applications
Edited by Grant Harper

International Standard Book Number: 978-1-68285-133-3 (Hardback)

Printed in the United States of America.

Contents

Preface

In an age where gadgets and technology go hand in hand with day to day activities of human beings, concepts such as mobile opportunistic networks bring the world closer. Research and advancements that will redefine the discipline of mobile opportunistic networks have been presented in this book. Video streaming, performance evaluation, architecture, protocols, algorithms, etc. are some of the concepts and applications that have been covered in this book. Scientists and students actively engaged in this discipline will find this book full of crucial and unexplored concepts.

This book is a comprehensive compilation of works of different researchers from varied parts of the world. It includes valuable experiences of the researchers with the sole objective of providing the readers (learners) with a proper knowledge of the concerned field. This book will be beneficial in evoking inspiration and enhancing the knowledge of the interested readers.

In the end, I would like to extend my heartiest thanks to the authors who worked with great determination on their chapters. I also appreciate the publisher's support in the course of the book. I would also like to deeply acknowledge my family who stood by me as a source of inspiration during the project.

Editor

VIDEO STREAMING PERFORMANCE UNDER WELL-KNOWN PACKET SCHEDULING ALGORITHMS

[1]Huda Adibah Mohd Ramli, [2]Kumbesan Sandrasegaran, [3]Riyaj Basukala, [4]Rachod Patachaianand and [5]Toyoba Sohana Afrin

[1,2,3,4,5]Faculty of Engineering and Information Technology, University of Technology, Sydney, Australia

[[1]HudaAdibah.MohdRamli, [2]kumbes, [4]rpatacha]@eng.uts.edu.au
[3]Riyaj.Basukala@uts.edu.au
[5]ToyobaSohana.Afrin@student.uts.edu.au
[1]International Islamic University Malaysia (IIUM), Kuala Lumpur, Malaysia

ABSTRACT

Video streaming is becoming increasingly popular among the wireless users. However, supporting video streaming over the wireless networks is not an easy task due to the dynamic radio propagation environment, limited radio resources as well as Quality of Service (QoS) requirements of the video streaming that need to be satisfied at acceptable levels. Most studies proposed to support video streaming are computationally expensive to be used in Orthogonal Frequency Division Multiple Access (OFDMA) based wireless IP networks. This paper evaluates video streaming performance under three well-known algorithms that are more practical to be used in the OFDMA based wireless IP networks due to their reduced complexity. It is demonstrated via computer simulation that Proportional Fair (PF) algorithm outperforms other well-known algorithms by providing video streaming QoS at acceptable levels whilst maximizing cell throughput.

KEYWORDS

Video streaming, Wireless IP Network, Packet scheduling, OFDMA, LTE

1. INTRODUCTION

Significant progresses in both wireless networking and video coding technologies have enabled video streaming to be supported over the wireless networks. An actual video stream contains data of a large size. The video stream is encoded into a smaller variable bit rate stream based on Moving Picture Experts Group (MPEG) or H.26x standards to reduce the bandwidth requirements [1]. The MPEG and H.26x are the two prominent standards developed by International Organization for Standardization/Information Centre (ISO/IEC) and International Telecommunication Union (ITU) standardization bodies.

There are multiple Groups of Pictures (GoP) within an encoded video stream. A sequence of frames starting with an I frame and all frames prior to the subsequent I frame constitutes a GoP [2]. An I frame represents a fully specified picture. It functions as a reference frame within a GoP. It is encoded without reference to any other frames except to itself [3]. A P frame contains only changes in the information from the previous frame. It is encoded by predicting the changes in the information in the previous I or P frame within its GoP. On the other hand, a B frame is encoded with reference to both previous and sub-sequent I or P frames. It contains the changes in the information from previous and subsequent frames.

Each frame within a GoP has different importance and is highly co-related due to the encoding. Therefore, it can be concluded that the I frame has the highest importance followed by the P and B frames. A lost of a reference frame within a GoP results in the lost of other dependent frames within the same GoP [4].

OFDMA is a multi-carrier access technology that divides the wide available bandwidth into multiple equally spaced and mutually orthogonal sub-carriers [5],[6]. Due to its robustness to inter-symbol interference and immunity to frequency selective fading [7], the OFDMA technology has been considered as a promising candidate for future wireless IP networks. Packet scheduling is one of the key RRM functions within the OFDMA based wireless IP networks for efficient radio resource utilizations and satisfying the QoS requirements of multimedia users. Moreover, the OFDMA technology allows packet scheduling to be performed in both time and frequency domains, thus enabling multi-user diversity to be exploited rather than combated.

Given that the video streaming service is becoming increasingly popular, a good portion of the radio resources needs to be allocated to provide support for the video users. However, majority of the packet scheduling studies that are proposed to support the video streaming are computationally expensive to be used in the OFDMA based wireless IP networks. Therefore, this paper evaluates video streaming performance under three well-known algorithms that are more practical to be used in the OFDMA based wireless IP networks due to their reduced complexity. This paper contributes to the identification of a well-known algorithm that is best suited for video streaming over the OFDMA based wireless IP networks.

The rest of this paper is organized as follows: Section 2 reviews on the related studies followed by detailed descriptions of the system model in Section 3. Discussions on the well-known packet scheduling algorithms and metrics used for performance evaluation are discussed in Section 4 and Section 5 respectively. Section 6 contains environment and results of the performance evaluation whereas conclusions are summarized in Section 7.

2. RELATED STUDIES

Adapting the wireless networks that have dynamic radio propagation environment and limited radio resources to satisfy the QoS requirements of video users at acceptable levels remains a challenge that needs to be addressed when supporting the video streaming over the OFDMA based wireless IP networks [8-11]. A number of packet scheduling studies have been proposed in the literature to address this challenge. The authors in [12-16] considered Channel Quality Information (CQI), average throughput and packet delay information of each user when scheduling video packets over the OFDMA based wireless networks. It is assumed in these studies that each video frame has equal importance, which is not typically the case in a realistic scenario.

To satisfy the QoS requirements of video users, the studies in [8, 17-19] considered video contents, commonly referred as content-aware, in their proposed algorithms. The proposed algorithms prioritize a video user on the basis of frame importance/deadline and its contribution towards improvement of the perceptual video quality. These studies assumed that a deadline is associated to each frame. If a frame arrives after the deadline, the frame and all other dependent frames are discarded. The overall goals of these studies are to improve perceptual video quality (minimize distortion) while minimizing the number of lost frames.

A more realistic video streaming scenario that utilized a playback buffer at the user is modelled in [20-23]. Under this scenario, the start of video playback is delayed until the total number of frames within the playback buffer exceeds a buffering threshold. This delay is referred as initial buffering delay and it aims to ensure a continuous video playback. The video and the wireless networks characteristics may lead to interruptions during the video playback. The video playback resumes at the same position where the interruption occurs after re-buffering takes place. Re-buffering is a process that re-fills the playback buffer into the specified buffering threshold.

The packet scheduling algorithms proposed to support video streaming under this realistic scenario [20-22] take video contents into consideration, which are similar to the studies in [8, 17-19]. However, on the contrary to the [8, 17-19] that mainly focused on satisfying the QoS requirements of the video users, the studies in [20-22] attempted to find a trade-off between maximizing throughput/capacity and satisfying the QoS requirements of video users.

There is always a trade-off between performance gain and complexity. A content-aware packet scheduling algorithm in the OFDMA based wireless IP networks will be computationally expensive with increasing number of users and the number of available radio resources [20], therefore not preferable from an implementation point of view [24]. In order to reduce complexity, this paper evaluates the video streaming performance under three well-known algorithms (i.e. Maximum-Rate (Max-Rate) [25], Round Robin (RR) [26] and PF [27]) over an OFDMA based wireless IP network. These algorithms are content un-aware packet scheduling algorithms which have been proposed for use in single-carrier wireless networks.

3. SYSTEM MODEL AND DESCRIPTIONS

A downlink Long Term Evolution (LTE) system consisting of a base station and N video streaming users are considered to represent an OFDMA based wireless IP network. It is assumed in this paper that (i) the serving base station buffers video streams that are received from the streaming server through a lossless and a high bandwidth backbone network [28]; (ii) each video frame of a variable size is segmented into groups of packets of a fixed size and (iii) all packets belonging to a frame need to be completely received for successful decoding at a user. The radio resource that is available to be shared by all users in the downlink LTE system is known as a Resource Block (RB). A RB is made up of 12 sub-carriers (180 kHz total bandwidth) of 1 ms duration [29],[30]. Moreover, a transmission time interval (TTI) of 1 ms duration is defined for packet scheduling in the downlink LTE system.

A user that has data to receive estimates its instantaneous Signal-to-Interference-Noise-Ratio (SINR) on each RB, converts it into a CQI value with the target block error rate (BLER) to be less than 10 % [29] and reports each CQI value to the serving base station. It is assumed in this paper that the CQI reporting is performed in each TTI and on each RB. However, these CQI values are available to be used by the packet scheduler after a certain delay.

The packet scheduler selects a user to receive its packets based on a packet scheduling algorithm on each RB and in each TTI. Note that a RB may only be allocated to a single user in a TTI but a user may be allocated variable number of RBs in the TTI. The packets to be transmitted to a user in a TTI are packed into one Transport Block (TB). The size of a TB depends on the supportable Modulation and Coding Schemes (MCSs) on each RB that is allocated to the user. The MCS is determined from the CQI value reported by the user.

Data transmissions over the wireless networks are prone to error. Therefore, the LTE system uses a stop-and-wait Incremental Redundancy Hybrid Automatic Repeat Request (IR HARQ) technique to recover wireless transmission errors. This is done by multiple retransmissions and soft combining of a corrupted TB. This leads to an improvement in the decoding efficiency at a user's soft buffer. If a TB of a user is corrupted during transmission, the user sends a Negative Acknowledgement (NACK) to the packet scheduler indicating retransmission of the TB is required. The packet scheduler placed all packets belonging to the corrupted TB at the front of the user's buffer. This indicates that the packets of a user that require retransmission are given a higher priority compared to its packets that are waiting for the first transmission. The packet scheduler drops all packets that have exceeded maximum number of retransmission, contrary to [21] that retransmit these packets until they are correctly received.

All packets that arrive at a user are stored into the playback buffer, which resides at the user's application layer, and re-constructed for video playback. These packets are packed into their associated frame. Some packets belonging to a frame may be lost. Therefore, all packets belonging to the frame are dropped and the frame is considered as a lost frame. Each video frame is decoded before they can be played back. Some frames cannot be decoded if a reference frame is lost. These frames are dropped at the user. The video frames are played back in sequence and at a fixed rate. If a frame within a sequence is lost, a simple error concealment method is used to estimate the missing information associated to the lost frame. This is done by duplicating the information of the latest successfully decoded frame into the position where the lost frame occurs [28]. The video frames are being played back in sequence and at a fixed rate.

4. WELL-KNOWN PACKET SCHEDULING ALGORITHMS

The Max-Rate, RR and PF are the well-known packet scheduling algorithms that are more likely to be used in the OFDMA based wireless IP networks due to their reduced complexity. These algorithms were proposed to support best effort services in the single-carrier wireless networks and aimed to maximize throughput and satisfy fairness of these services.

The Max-Rate is an algorithm that always selects a user having a good channel quality. On the other hand, a user having a poor channel quality is less likely to be given any opportunity to receive its packets unless its channel quality improves. The Max-Rate algorithm is a good candidate for throughput maximization but is not capable of satisfying fairness between the users. The mathematical expression for the Max-Rate algorithm is given as follows:

$$k = \arg \max r_i(t) \quad i \in N \tag{1}$$

where $r_i(t)$ is the total number of supportable bits (based on MCS) of user i at time interval t and N is the total number of users.

Given that fairness has been an issue in the Max-Rate algorithm, the RR algorithm is proposed to address this situation. The RR algorithm gives equal opportunity to each user to receive its packets in a cyclic fashion. Though the RR algorithm is able to considerably improve fairness performance between the users, the throughput performance is significantly degraded.

The PF algorithm provides a better trade-off between throughput maximization and fairness satisfaction. This algorithm is originally proposed to support non-real time services in Code Division Multiple Access (CDMA) system. The PF algorithm selects user k to receive its packets at time interval t based on the following equations:

$$k = \arg \max \frac{r_i(t)}{R_i(t)} \quad i \in N \tag{2}$$

$$R_i(t) = \left(1 - \frac{1}{t_c}\right) R_i(t-1) + \frac{1}{t_c} * r_i(t-1) \quad i \in N \tag{3}$$

where $r_i(t)$ and $R_i(t)$ are the total number of supportable bits (based on MCS) and the average throughput of user i at time interval t, t_c is a time constant and N is the total number of users.

The t_c value determines between maximizing throughput and satisfying fairness in the PF algorithm. The PF algorithm behaves similar to the Max-Rate algorithm if a higher t_c value is used while similar to the RR algorithm for a lower value of t_c. A t_c value of 1000 ms as proposed in [27] that can provide a trade-off between throughput and fairness is used in this paper.

Packet scheduling in the single-carrier wireless networks allocates all the available radio resources to a single user in each scheduling interval. Therefore, some modifications are made to enable these well-known algorithms to support packet scheduling in the OFDMA based wireless IP networks (specifically focusing on the downlink LTE system). Since there are multiple RBs available to be shared by the users in each TTI, these algorithms select user k to receive its packet on RB j based on equation (4) for Max-Rate and equations (5-7) for PF. Similarly the RR algorithm is modified such that a user that is to receive its packet on RB j is selected in a cyclic fashion.

$$k = \arg\max r_{i,j}(t) \quad i \in N \tag{4}$$

$$k = \arg\max \frac{r_{i,j}(t)}{R_i(t)} \quad i \in N \tag{5}$$

$$R_i(t) = \left(1 - \frac{1}{t_c}\right)R_i(t-1) + \frac{1}{t_c} * rtot_i(t-1) \quad i \in N \tag{6}$$

$$rtot_i(t-1) = \sum_{j=1}^{\max RB} r_{i,j}(t-1) * \Delta_{i,j}(t-1) \quad i \in N \tag{7}$$

where $r_{i,j}(t)$ is the total number of supportable bits (based on MCS) on RB j of user i at time interval t, $R_i(t)$ is the average throughput of user i at time interval t, $rtot_i(t)$ is the total number of supportable bits on the RBs allocated to user i at time interval $t-1$, t_c is a time constant, N is the total number of users and *max RB* is the total number of RBs. The variable $\Delta_{i,j}(t-1)$ contains a value between 0 and 1 in which $\Delta_{i,j}(t-1)=1$ if RB j is selected for user i at time interval $t-1$ and 0 otherwise.

5. PERFORMANCE METRICS

The video streaming performance under the three well-known algorithms is evaluated on the basis of minimum Mean Opinion Score (MOS), Freezing Delay Ratio (FDR) and cell throughput. The MOS and FDR are the metrics that are related to the QoS requirements of video streaming whereas the cell throughput is a network related performance metric. Detailed descriptions of each metric are provided as follows:

The minimum MOS represents both the perceptual video quality (measured in terms of Peak-Signal-to-Noise Ratio, PSNR [31]) as well as the fairness experienced among all video users. The PSNR gives the difference in terms of peak signal between the source and the re-constructed video frame (i.e. the source video frame is the encoded frame located at the serving base station before it is streamed to the user). It has the following mathematical expression [32]:

$$PSNR_{i,m} = 20\log 10 \frac{255}{D_{i,m}} \qquad (8)$$

$$D_{i,m} = \sqrt{\frac{1}{XY} \sum_{x=0}^{X-1} \sum_{y=0}^{Y-1} \left[F_{i,m}(x, y) - f_{i,m}(x, y) \right]^2} \qquad (9)$$

where $PSNR_{i,m}$ is the PSNR value at the mth frame of user i, X and Y are resolution in pixels, $F_{i,m}(x,y)$ and $f_{i,m}(x,y)$ are the source and reconstructed pixel's values at the mth video frame at position (x, y) of user i.

It is assumed that each user requests a total of F frames throughout a video session. A video session starts at the beginning of the simulation ($t=1$ ms) and continues until the total simulation time ($t=T$ ms). The total number of frames that are able to be played back throughout the video session varies among the users. The mean PSNR of each user (averaged over F frames) is determined as follows:

$$mean\ PSNR_i = \frac{1}{F_i} \sum_{m=1}^{pf_i} PSNR_{i,m} \qquad (10)$$

where $PSNR_{i,m}$ is the PSNR value at the mth frame of user i, $mean\ PSNR_i$ and F_i are the mean PSNR and the total number of frames requested by user i and pf_i is the frame number of user i that is currently being played back when the session ends.

The minimum MOS, as given in Table 1, represents a user having the lowest mean PSNR among other users. It ensures that equal perceptual video quality is experienced among the video users. A low minimum MOS indicates that at least one user is not satisfied whereas maximization of the minimum MOS ensures fairness and high perceptual video quality are experienced among the video users.

Table 1. Mean PSNR to MOS Mapping [33]

PSNR (dB)	MOS Value	Quality
< 20	1	Bad
20 – 25	2	Poor
25 – 31	3	Fair
31 – 37	4	Good
> 37	5	Excellent

A user will not be satisfied if it has to wait longer to start or resume its video playback. Therefore, it is necessary to minimize the FDR throughout the video session. The freezing delay, as discussed in [23], is the total delay in the video playback due to (i) scheduling and propagation delay (ε); (ii) initial buffering delay (α) and (iii) re-buffering delay (β).

An example of freezing delay experiences by a user throughout a video session is illustrated in Figure 1. It is shown in the figure that, after a request for a video stream is made by the user, it takes ε ms for the first bit of the video stream to arrive and being stored into the user's playback buffer. The video playback starts after frames of α ms duration are available within the playback buffer (i.e. total duration of frames equals or greater than the buffering threshold). The total duration of frames within the playback buffer (T_{pb}) can be determined as follows:

$$T_{pb} = \frac{F_{pb}}{\Omega} \tag{11}$$

where Ω is the frame rate (in fps) and F_{pb} is the total number of frames within the playback buffer.

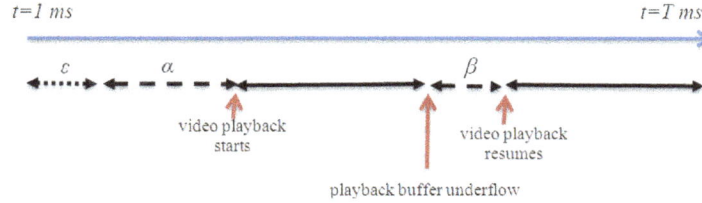

Figure 1. Example of freezing delay during a video session.

Video playback is interrupted and re-buffering takes place if the total duration of frames within the playback buffer falls below $\alpha - \beta$ ms. The FDR is defined in this paper as the mean of the ratio of total freezing delay of each user to the total simulation time as given as folows:

$$FDR = \frac{1}{N} \sum_{i=1}^{N} \frac{Df_i}{T} \tag{12}$$

where Df_i is the total freezing delay of user i, T is the total simulation time and N is the total number of users.

Finally, the cell throughput is defined as the total number of bits correctly received by all users per second. The cell throughput is measured at the base station. It is mathematically expressed as:

$$cell\ throughput = \frac{1}{T} \sum_{i=1}^{N} \sum_{t=1}^{T} tput_i(t) \tag{13}$$

where $tput_i(t)$ is the total size of correctly received packets (in bits) of user i at time interval t, T is the total simulation time and N is the total number of users.

6. SIMULATION ENVIRONMENTS AND RESULTS

A computer simulation using C++ platform is used to evaluate the minimum MOS, FDR and cell throughput performances of the three well-known algorithms for varying number of users and buffering thresholds. The environments and results of the computer simulation are discussed in the following sub-sections.

6.1. Simulation Environment

A single hexagonal cell scenario of 5 MHz bandwidth with 25 RBs and 2 GHz carrier frequency is modelled. The base station has a fixed location at the centre cell and it is assumed that equal transmit power (43.01 dBm total base station transmit power) is used on each RB. The cell contains 5 – 20 video streaming users that are moving within the cell radius of 200 m at a constant speed of 10 km/h in random directions. These users are uniformly located within the cell. A user is wrapped-around whenever it reaches the cell edge [34]. The Cost-231 HATA model for an urban environment [35], a Gaussian log-normal distribution with 0 mean and 8 dB standard deviation [36] and a frequency-flat Rayleigh fading model [37] are used to model the radio propagation channel. A total of 16 CQI levels as defined in [29] are used. The CQI and HARQ reporting are modeled error free with 3 ms CQI delay and 4 ms HARQ delay. The maximum number of retransmissions is limited to 3 and the threshold for re-sequencing timer is set to 500 ms.

Three video streams in a 352x288 resolution which are downloaded from a publicly available video traces [2] are used. Video trace is a simple file that contains number, type/importance, decoding/video playback deadline, size and quality of each video frame [38, 39]. The video streams are encoded with a frame rate of 30 fps (frames per second). One GoP contains 16 frames with the IBBBPBBBPBBBPBBB sequence. A user may only request one random video stream throughout its session and the request arrives at the beginning of the simulation. The requested video stream may start at any frame number as long as it is an I frame. Only 300 frames out of the total frames are simulated for each user. The maximum size of the playback buffer is infinite. The video playback is interrupted if there are less than 5 frames remaining within the playback buffer. The buffering thresholds vary between 266.667 ms – 800 ms duration [18].

6.2 Results –Vary Number of Users

This sub-section evaluates the impacts of varying number of users towards minimum MOS, FDR and cell throughput performances under the Max-Rate, RR and PF algorithms. The buffering threshold of each user is fixed at 800 ms. Figure 2 shows the minimum MOS with increasing number of users. As expected the minimum MOS degrades with increasing number of users. The figure shows that the PF algorithm is able to maintain a good perceptual video quality and satisfy fairness for a higher number of users compared to the Max-Rate and RR algorithms. When the number of users equals to 10, the PF algorithm is able to maintain a good perceptual video quality (minimum MOS=4) for all 10 users. However, at this number of users, at least one user is experiencing a fair (minimum MOS=3) and a poor (minimum MOS=2) perceptual video quality under the RR and Max-Rate algorithms respectively.

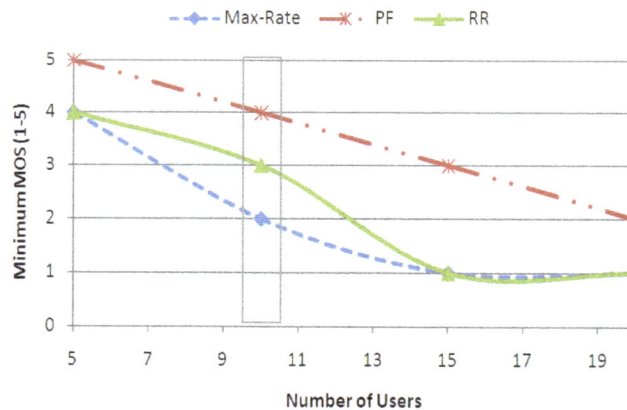

Figure 2. Minimum MOS vs. number of users

The FDR performance with increasing number of users is shown in Figure 3. The FDR degrades with increasing number of users and it is demonstrated within the figure that the PF is having the best FDR performance followed by the RR and Max-Rate algorithms. The worst FDR performance under the Max-Rate algorithm is mostly due to the users having poor channel qualities that have to wait longer before they can start or resume their video playbacks.

Moreover, over-allocation of the available RBs to the users with good channel qualities does not benefit the Max-Rate algorithm as the video is played back at a fixed rate.

The RR algorithm that allocates equal share of transmission time to all users achieves a better FDR performance compared to the Max-Rate algorithm. However, it is not possible for this algorithm to minimize the freezing delay among the users as the channel quality of each user is different. The PF algorithm achieves the best FDR performance for selecting users having relatively good channel qualities whilst ensuring that all users receive equal amount of packets based on their average throughput update (denominator in (5)). This minimizes the delays taken by the users to start or resume their video playbacks.

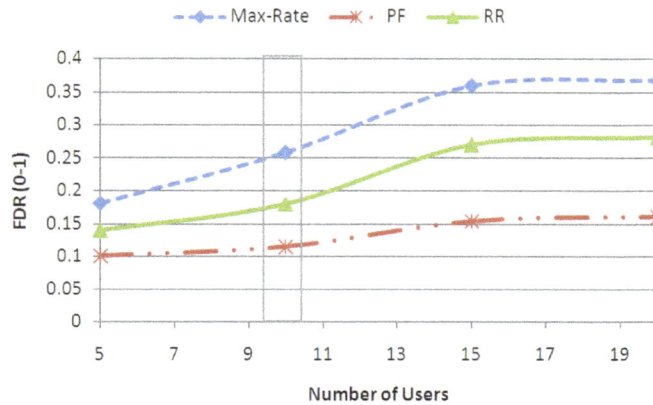

Figure 3. FDR vs. number of users

The cell throughput performance under the three well-known algorithms with increasing number of users is shown in Figure 4. With increasing number of users, the cell throughput achieves under the PF outperforms the Max-Rate and RR algorithms. The PF algorithm always selects the users having relatively good channel qualities for transmissions. By delaying transmissions of some users, there is a possibility that the channel qualities of these users improve over time. This allows the users to receive their packets on the RBs with better channel qualities, thus improving the cell throughput under the PF algorithm.

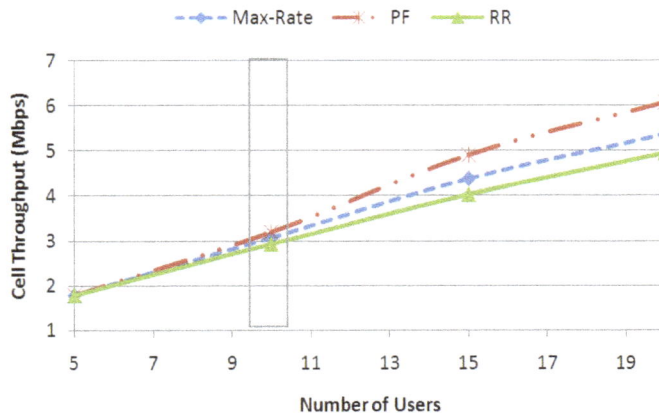

Figure 4. Cell throughput vs. number of users

6.3 Results –Vary Buffering Thresholds

The impacts of varying buffering thresholds while setting the number of users fixed at 10 users are evaluated in this sub-section. The minimum MOS under the three well-known algorithms is shown in Figure 5. The figure demonstrates that the minimum MOS under the RR and PF algorithms remain the same within the buffering thresholds ranging from 266.667 – 800 ms. However, the minimum MOS under the Max-Rate algorithm degrades when the buffering threshold is set to 800 ms. The Max-Rate algorithm deprives the users having poor channel qualities from receiving their packets. With increasing buffering thresholds, only small number of the requested frames may be able to be played back since the users may have to wait longer to start or resume their video playbacks. This leads to the degradation of the minimum MOS at a higher buffering threshold.

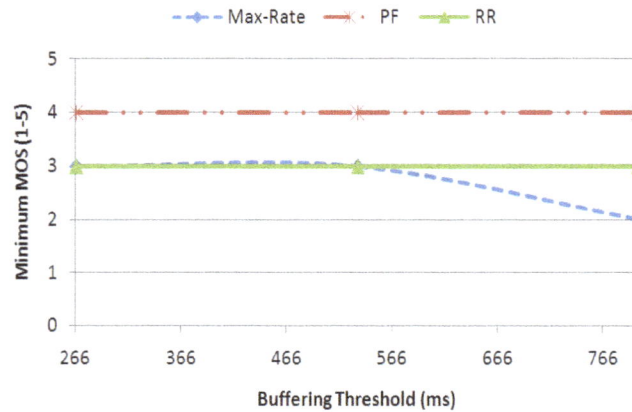

Figure 5. Minimum MOS vs buffering threshold

Figure 6 shows the FDR performance under the three well-known algorithms. With increasing buffering thresholds, the PF algorithm is having the best FDR performance followed by the RR and Max-Rate algorithms. Moreover, Table 2 shows that, the FDR performances under all algorithms slightly degrade with increasing buffering thresholds.

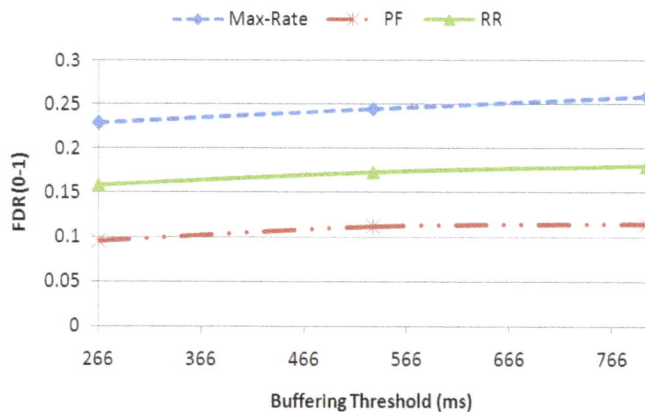

Figure 6. FDR vs buffering threshold

Table 2. Percentage of FDR degradation

	% of FDR Degradation (266.667 – 533.33)	*% of FDR Degradation (533.33 – 800)*	*% of FDR Degradation (Mean)*
Max-Rate	6.910835	5.4323	6.171568
PF	16.57972	2.622393	9.601055
RR	9.757021	3.657064	6.707043

The buffering of video frames takes place at a user's application layer. Therefore, varying the buffering thresholds does not have any impact towards the cell throughput performance (as shown in Figure 7) which is a metric being determined at the physical layer of the base station.

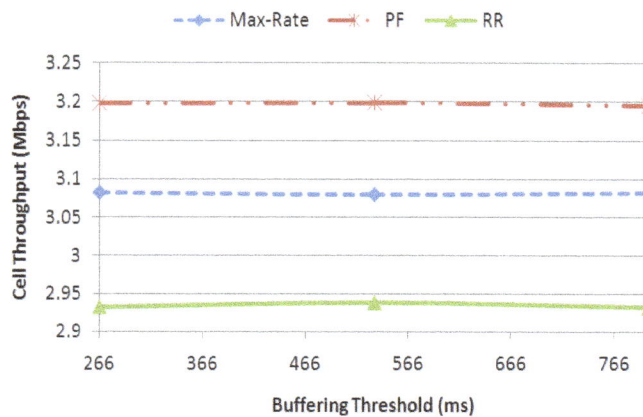

Figure 7. Cell throughput vs buffering threshold

7. CONCLUSIONS

This paper evaluates video streaming performance on the basis of minimum MOS, FDR and cell throughput under Max-Rate, PF and RR algorithms over an OFDMA based wireless IP network. These well-known algorithms have a reduced complexity compared to the content-aware packet scheduling algorithms. A more realistic video streaming scenario and performance metrics are used in the performance evaluations. It is demonstrated in the simulation results that the PF algorithm is able to provide the QoS requirements at acceptable levels for a higher number of video users compared to the Max-Rate and RR algorithms. Moreover, it is also demonstrated in the simulation results that the buffering thresholds in the range of 266.667 ms to 800 ms do not significantly impact the FDR and cell throughput performances under the three well-known algorithms when the number of users is small.

REFERENCES

[1]   H. Sun, A. Vetro, and J. Xin, "An Overview of Scalable Video Streaming," *Wireless Communications and Mobile Computing,* vol. 7, no. 2, pp. 159-172, February 2007.

[2]   P. Seeling, M. Reisslein, and B. Kulapala, "Network Performance Evaluation Using Frame Size and Quality Traces of Single-Layer and Two-Layer Video: A Tutorial," *IEEE Communications Surveys & Tutorials,* vol. 6, no. 3, pp. 58-78, 2004.

[3]   Wikipedia, "Video Compression Picture Types [Online] Available: http://en.wikipedia.org/wiki/Video_compression_picture_types," Accessed: Oct. 23, 2010.

[4]   E. Haghani, P. Shyam, C. Doru, K. Eunyoung, and N. Ansari, "A Quality-Driven Cross-Layer Solution for MPEG Video Streaming Over WiMAX Networks," *IEEE Transactions on Multimedia*, vol. 11, no. 6, pp. 1140-1147, 2009.

[5]   3GPP, 'Evolved Universal Terrestrial Radio Access (E-UTRA) and Evolved Universal Terrestrial Radio Access Network (E-UTRAN); Overall Description; Stage 2 (Release 9)," TS 36.300, version 9.2.0, December 2009.

[6]   A. Omri, R. Bouallegue, R. Hamila, and M. Hasna, "Channel Estimation for LTE Uplink System by Perceptron Neural Network," *International Journal of Wireless & Mobile Networks (IJWMN)*, vol. 2, no. 3, pp. 155-165, 2010.

[7]   S. Nonchev and M. Valkama, "Advanced Radio Resource Management for Multiantenna Packet Radio Systems," *International Journal of Wireless & Mobile Networks (IJWMN)*, vol. 2, no. 2, pp. 1-14, 2010.

[8]   Z. Honghai, Z. Yanyan, M. A. Khojastepour, and S. Rangarajan, "Scalable Video Streaming over Fading Wireless Channels," in *IEEE Wireless Communications and Networking Conference*, 2009, pp. 1-6.

[9]   E. Maani, P. V. Pahalawatta, R. Berry, T. N. Pappas, and A. K. Katsaggelos, "Resource Allocation for Downlink Multiuser Video Transmission Over Wireless Lossy Networks," *IEEE Transactions on Image Processing*, vol. 17, no. 9, pp. 1663-1671, 2008.

[10]   W. Dalei, L. Haiyan, C. Song, W. Haohong, and A. Katsaggelos, "Quality-Driven Optimization for Content-Aware Real-Time Video Streaming in Wireless Mesh Networks," in *IEEE Global Telecommunications Conference*, 2008, pp. 1-5.

[11]   J. Sengupta and G. S. Grewal, "Performance Evaluation of IEEE 802.11 MAC Layer in Supporting Delay Sensitive Services," *International Journal of Wireless & Mobile Networks (IJWMN)*, vol. 2, no. 1, pp. 42-53, 2010.

[12]   L. Xiantao, L. Guangyi, W. Ying, and Z. Ping, "Downlink Packet Scheduling for Real-Time Traffic in Multi-User OFDMA System," in *IEEE 64th Vehicular Technology Conference*, 2006, pp. 1 - 5

[13]   S. Shin and B.-H. Ryu, "Packet Loss Fair Scheduling Scheme for Real-Time Traffic in OFDMA Systems," *ETRI*, vol. 26, October 2004.

[14]   Q. Sun, H. Tian, K. Dong, R. Zhou, and P. Zhang, "Packet Scheduling for Real-Time Traffic for Multiuser Downlink MIMO-OFDMA Systems," in *IEEE Wireless Communications and Networking Conference*, 2008, pp. 1849-1853.

[15]   K. Sandrasegaran, H. A. M. Ramli, and R. Basukala, "Delay-Prioritized Scheduling (DPS) for Real Time Traffic in 3GPP LTE System," in *IEEE Wireless Communications and Networking Conference* 2010, pp. 1-6.

[16]   H. A. M. Ramli, K. Sandrasegaran, R. Basukala, R. Patachaianand, M. Xue, and C.-C. Lin, "Resource Allocation Technique for Video Streaming Applications in the LTE System," in *19th Annual Wireless and Optical Communications Conference (WOCC)*, 2010, pp. 1-5.

[17]   G. Liebl, T. Schierl, T. Wiegand, and T. Stockhammer, "Advanced Wireless Multiuser Video Streaming using the Scalable Video Coding Extensions of H.264/MPEG4-AVC," in *IEEE International Conference on Multimedia and Expo*, 2006, pp. 625-628.

[18]   J. Xin, H. Jianwei, C. Mung, and F. Catthoor, "Downlink OFDM Scheduling and Resource Allocation for Delay Constraint SVC Streaming," in *IEEE International Conference on Communications*, 2008, pp. 2512-2518.

[19]   T. Junhua, Z. Liren, and S. Chee-Kheong, "An Opportunistic Scheduling Algorithm for MPEG Video Over Shared Wireless Downlink," in *IEEE International Conference on Communications*. vol. 2, 2006, pp. 872-877.

[20]   P. E. Omiyi and M. G. Martini, "Cross-Layer Content/Channel Aware Multi-User Scheduling for Downlink Wireless Video Streaming," in *5th IEEE International Symposium on Wireless Pervasive Computing (ISWPC)*, 2010, pp. 412-417.

[21]   T. Ozcelebi, M. Oguz Sunay, A. Murat Tekalp, and M. Reha Civanlar, "Cross-Layer Optimized Rate Adaptation and Scheduling for Multiple-User Wireless Video Streaming," *IEEE Journal on Selected Areas in Communications*, vol. 25, no. 4, pp. 760-769, 2007.

[22]   C. Yongjin, C. C. J. Kuo, H. Renxiang, and C. Lima, "Cross-Layer Design for Wireless Video Streaming," in *IEEE Global Telecommunications Conference*, 2009, pp. 1-5.

[23]   V. Vukadinovic and G. Karlsson, "Video Streaming Performance under Proportional Fair Scheduling," *IEEE Journal on Selected Areas in Communications*, vol. 28, no. 3, pp. 399-408, 2010.

[24]   A. K. F. Khattab and K. M. F. Elsayed, "Opportunistic Scheduling of Delay Sensitive Traffic in OFDMA-based Wireless Networks," in *International Symposium on a World of Wireless, Mobile and Multimedia Networks*, 2006, pp. 288-297.

[25]   B. S. Tsybakov, "File Transmission over Wireless Fast Fading Downlink," *IEEE Transactions on Information Theory,* vol. 48, no. 8, pp. 2323-2337, 2002.

[26]   E. Dahlman, S. Parkvall, J. Skold, and P. Beming, *3G Evolution: HSPA and LTE for Mobile Broadband*, First ed.: Elsevier Ltd., 2007.

[27]   A. Jalali, R. Padovani, and R. Pankaj, "Data Throughput of CDMA-HDR a High Efficiency-High Data Rate Personal Communication Wireless System," in *IEEE 51st Vehicular Technology Conference Proceedings*. vol. 3, 2000, pp. 1854-1858.

[28]   L. Fan, L. Guizhong, and H. Lijun, "Application-Driven Cross-Layer Approaches to Video Transmission over Downlink OFDMA Networks," in *IEEE GLOBECOM Workshops*, 2009, pp. 1-6.

[29]   H. Holma and A. Toskala, *LTE for UMTS: OFDMA and SC-FDMA Based Radio Access*, 1 ed.: John Wiley & Sons Ltd., 2009.

[30]   3GPP, "Physical Layer Procedures (FDD) (Release 9)," TS 25.214, version 9.1.0, December 2009.

[31]   B. Pathak, H., G. Childs, and A. Maaruf, "Implementation of NEWPRED and UEP for Video Quality Improvements for MPEG-4 over Simulated UMTS Channel Propagation Environments," *International Journal of Wireless & Mobile Networks (IJWMN)*, vol. 2, no. 2, pp. 94-108, 2010.

[32]   P. Seeling, M. Reisslein, and F. H. P. Fitzek, "Layered Video Coding Offset Distortion Traces for Trace-Based Evaluation of Video Quality after Network Transport," in *3rd IEEE Consumer Communications and Networking Conference*. vol. 1, 2006, pp. 292-296.

[33]   J. Klaue, B. Rathke, and A. Wolisz, "Evalvid - A Framework for Video Transmission and Quality Evaluation," in *13th International Conference on Modelling Techniques and Tools for Performance Evaluation* Urbana, Illinois, USA, 2003, pp. 1-5.

[34]   A. G. Orozco Lugo, F. A. Cruz Prez, and G. Hernandez Valdez, "Investigating the Boundary Effect of a Multimedia TDMA Personal Mobile Communication Network Simulation," in *IEEE VTS 54th Vehicular Technology Conference*. vol. 4, 2001, pp. 2740-2744.

[35]   T. S. Rappaport, *Wireless Communications: Principles and Practice*, 3rd ed.: Prentice Hall PTR, pp. 153-154, 2003.

[36]   M. Gudmundson, "Correlation Model for Shadow Fading in Mobile Radio Systems," in *Electronics Letters*. vol. 27, 1991, pp. 2145-2146.

[37]   C. Komninakis, "A Fast and Accurate Rayleigh Fading Simulator," in *IEEE Globecom* San Francisco, CA, 2003, pp. 3306-3310.

[38]   M. Reisslein, L. Karam, and P. Seeling, "H.264/AVC and SVC Video Trace Library: A Quick Reference Guide," May, 2009.

[39]   G. V. D. Auwera and M. Reisslein, "Implications of Smoothing on Statistical Multiplexing of H.264/AVC and SVC Video Streams," *IEEE Transactions on Broadcasting,* vol. 55, no. 3, September 2009.

Authors

Huda Adibah Mohd Ramli is currently a PhD candidate in the Faculty of Engineering and Information Technology, University of Technology, Sydney (UTS), Australia. She received an M.Sc. in Software Engineering from University of Technology Malaysia (2006) and B.Eng. in Electrical and Computer Engineering from International Islamic University Malaysia (2003). Her current research interests focus on radio resource management for the future wireless IP networks.

Kumbesan Sandrasegaran (Sandy) holds a PhD in Electrical Engineering from McGill University (Canada) (1994), a Masters of Science Degree in Telecommunication Engineering from Essex University (UK) (1988) and a Bachelor of Science (Honours) Degree in Electrical Engineering (First Class) (UZ) (1985). He was a recipient of the Canadian Commonwealth Fellowship (1990-1994) and British Council Scholarship (1987-1988). He is a Professional Engineer (Pr.Eng) and has more than 20 years experience working either as a practitioner, researcher, consultant and educator in telecommunication networks.

During this time, he has focused on the planning, modeling, simulation, optimisation, security, and management of telecommunication networks.

Riyaj Basukala received his Masters of Engineering (Telecommunication Engineering) Degree from UTS in 2009 and Bachelors of Engineering in Electronics and Communication from Tribhuwan University, Nepal in 2005. Currently he is involved in working as a Research Assistant in Centre for Real Time Information Networks (CRIN) in UTS focussing mainly on developing simulation tools for simulating radio resource management in various radio interface technologies and also few other communication-based research projects.

Rachod Patachaianand is currently a doctoral candidate in the Faculty of Engineering and Information Technology, University of Technology Sydney (UTS), Australia. He received an M.Eng. in Telecommunications Engineering from UTS in 2007 and B.Eng. in Electrical Engineering from Chulalongkorn University, Thailand, in 2004. His current research has focussed on radio resource management of future mobile communication systems specifically focusing on link adaptation techniques with limited feedback.

Toyoba Sohana Afrin is currently a PhD candidate in the Faculty of Engineering and Information Technology, University of Technology, Sydney, Australia. She received a B.Sc Engineering in Electrical and Electronic Engineering from Bangladesh University of Engineering and Technology (2004). She gains 6 years working experiences in the field of Radio Access Network with telecom industry (2004-2010). Her current research explores in Radio Resource Management for future mobile communication systems.

2

PERFORMANCE IMPROVEMENT OF PAPR REDUCTION FOR OFDM SIGNAL IN LTE SYSTEM

Md. Munjure Mowla[1] and S.M. Mahmud Hasan[2]

[1]Department of Electrical & Electronic Engineering, Rajshahi University of Engineering & Technology, Rajshahi, Bangladesh

rimonece@gmail.com

[2] Department of Electronics & Telecommunication Engineering, Rajshahi University of Engineering & Technology, Rajshahi, Bangladesh

etemahmud@gmail.com

ABSTRACT

Orthogonal frequency division multiplexing (OFDM) is an emerging research field of wireless communication. It is one of the most proficient multi-carrier transmission techniques widely used today as broadband wired & wireless applications having several attributes such as provides greater immunity to multipath fading & impulse noise, eliminating inter symbol interference (ISI), inter carrier interference (ICI) & the need for equalizers. OFDM signals have a general problem of high peak to average power ratio (PAPR) which is defined as the ratio of the peak power to the average power of the OFDM signal. The drawback of high PAPR is that the dynamic range of the power amplifier (PA) and digital-to-analog converter (DAC). In this paper, an improved scheme of amplitude clipping & filtering method is proposed and implemented which shows the significant improvement in case of PAPR reduction while increasing slight BER compare to an existing method. Also, the comparative studies of different parameters will be covered.

KEYWORDS

Bit Error rate (BER), Complementary Cumulative Distribution Function (CCDF), Long Term Evolution (LTE), Orthogonal Frequency Division Multiplexing (OFDM) and Peak to Average Power Ratio (PAPR).

1. INTRODUCTION

Long term evolution (LTE) is the one of the latest steps toward the 4th generation (4G) of radio technologies designed to increase the capacity and speed of mobile telephone networks. It is identical by the third generation partnership project (3GPP) and is an evolution to existing 3G technologies in order to meet projected customer needs over the next decades. LTE has adopted orthogonal frequency division multiplexing (OFDM) for the transmission. OFDM meets the LTE requirement for spectrum flexibility and enables cost-efficient solutions for very wide carriers [1]. The supplementary increasing demand on high data rates in wireless communications systems has arisen in order to carry broadband services. OFDM offers high spectral efficiency, immune to the multipath fading, low inter-symbol interference (ISI), immunity to frequency selective fading and high power efficiency. Today, OFDM is used in many emerging fields like wired Asymmetric Digital Subscriber Line (ADSL), wireless Digital Audio Broadcast (DAB), wireless Digital Video Broadcast - Terrestrial (DVB - T), IEEE 802.11 Wireless Local Area Network (WLAN), IEEE

802.16 Broadband Wireless Access (BWA), Wireless Metropolitan Area Networks (IEEE 802.16d), European ETSI Hiperlan/2 etc [2].

One of the major problems of OFDM is high peak to average power ratio (PAPR) of the transmit signal. If the peak transmit power is limited by either regulatory or application constraints, the effect is to reduce the average power allowed under multicarrier transmission relative to that under constant power modulation techniques. This lessens the range of multicarrier transmission. Furthermore, the transmit power amplifier must be operated in its linear region (i.e., with a large input back-off), where the power conversion is inefficient to avoid spectral growth of the multicarrier signal in the form of intermodulation among subcarriers and out-of-band radiation. This may have a deleterious effect on battery lifetime in mobile applications. As handy devices have a finite battery life it is significant to find ways of reducing the PAPR allowing for a smaller more efficient high power amplifier (HPA), which in turn will mean a longer lasting battery life. In many low-cost applications, the problem of high PAPR may outweigh all the potential benefits of multicarrier transmission systems [3]. A number of promising approaches or processes have been proposed & implemented to reduce PAPR with the expense of increase transmit signal power, bit error rate (BER) & computational complexity and loss of data rate, etc. So, a system trade-off is required. These techniques include Amplitude Clipping and Filtering, Peak Windowing, Peak Cancellation, Peak Reduction Carrier, Envelope Scaling, Decision-Aided Reconstruction (DAR), Coding, Partial Transmit Sequence (PTS), Selective Mapping (SLM), Interleaving, Tone Reservation (TR), Tone Injection (TI), Active Constellation Extension (ACE), Clustered OFDM, Pilot Symbol Assisted Modulation, Nonlinear Companding Transforms etc [4].

2. CONCEPTUAL MODEL OF OFDM SYSTEM

OFDM is a special form of multicarrier modulation (MCM) with densely spaced subcarriers with overlapping spectra, thus allowing multiple-access [5]. MCM works on the principle of transmitting data by dividing the stream into several bit streams, each of which has a much lower bit rate and by using these sub-streams to modulate several carriers.

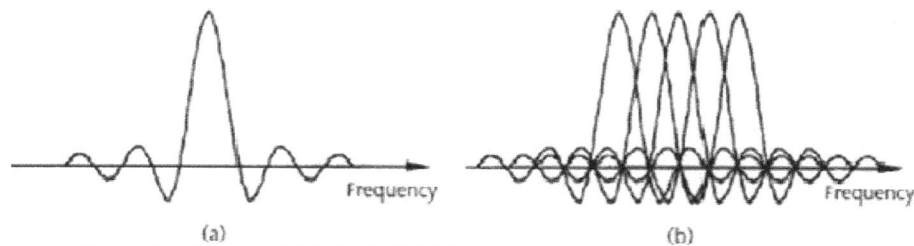

Figure 1.   Spectra of (a) An OFDM Sub-channel and (b) An OFDM Signal [2]

In multicarrier transmission, bandwidth divided in many non-overlapping subcarriers but not necessary that all subcarriers are orthogonal to each other as shown in figure 1 (a) [2]. In OFDM the sub-channels overlap each other to a certain extent as can be seen in figure 1 (b), which leads to an efficient use of the total bandwidth. The information sequence is mapped into symbols, which are distributed and sent over the N sub-channels, one symbol per channel. To permit dense packing and still ensure that a minimum of interference between the sub-channels is encountered, the carrier frequencies must be chosen carefully. By using orthogonal carriers, frequency domain can be viewed so as the frequency space between two sub-carriers is given by the distance to the first spectral null [2].

2.1. Mathematical Explanation of OFDM Signals

In OFDM systems, a defined number of successive input data samples are modulated first (e.g, PSK or QAM), and then jointly correlated together using inverse fast Fourier transform (IFFT) at the transmitter side. IFFT is used to produce orthogonal data subcarriers. Let, data block of length N is represented by a vector, $X=[X_0, X_{1......} X_{N-1}]^T$. Duration of any symbol X_K in the set X is T and represents one of the sub-carriers set. As the N sub-carriers chosen to transmit the signal are orthogonal, so we can have, $f_n = n\Delta f$, where $n\Delta f = 1/NT$ and NT is the duration of the OFDM data block X. The complex data block for the OFDM signal to be transmitted is given by [3],

$$x(t) = \frac{1}{\sqrt{N}} \sum_{n=0}^{N-1} X_n e^{j2\pi n\Delta ft} \qquad 0 \leq t \leq NT \qquad (1)$$

Where,

$j = \sqrt{-1}$, Δf is the subcarrier spacing and NT denotes the useful data block period. In OFDM the subcarriers are chosen to be orthogonal (i.e., $\Delta f = 1/NT$). However, OFDM output symbols typically have large dynamic envelope range due to the superposition process performed at the IFFT stage in the transmitter.

3. OVERVIEW OF PAPR

Presence of large number of independently modulated sub-carriers in an OFDM system the peak value of the system can be very high as compared to the average of the whole system. Coherent addition of N signals of same phase produces a peak which is N times the average signal [3]. PAPR is widely used to evaluate this variation of the output envelope. PAPR is an important factor in the design of both high power amplifier (PA) and digital-to-analog (D/A) converter, for generating error-free (minimum errors) transmitted OFDM symbols. So, the ratio of peak power to average power is known as PAPR.

$$PAPR = \frac{Peak\_Power}{Average\_Power}$$

The PAPR of the transmitted signal is defined as [6],

$$PAPR[x(t)] = \frac{\max_{0 \leq t \leq NT} |x(t)|^2}{P_{av}} = \frac{\max_{0 \leq t \leq NT} |x(t)|^2}{\frac{1}{NT} \int_0^{NT} |x(t)|^2 \, dt} \qquad (2)$$

Where, P_{av} is the average power of and it can be computed in the frequency domain because Inverse Fast Fourier Transform (IFFT) is a (scaled) unitary transformation.

To better estimated the PAPR of continuous time OFDM signals, the OFDM signals samples are obtained by L times oversampling [3]. L times oversampled time domain samples are LN point IFFT of the data block with $(L-1)N$ zero-padding. Therefore, the oversampled IFFT output can be expressed as,

$$x[n] = \frac{1}{\sqrt{N}} \sum_{k=0}^{N-1} X_k e^{j2\pi nk/LN} \qquad 0 \leq n \leq NL-1 \qquad (3)$$

It is known that the PAPR of the continuous-time signal cannot be obtained precisely by the use of Nyquist rate sampling, which corresponds to the case of L = 1. It is shown in that L = 4 can provide sufficiently accurate PAPR results.

The PAPR computed from the L-times oversampled time domain signal samples is given by,

$$PAPR\{x[n]\} = \frac{\max_{0 \leq t \leq NL-1} |x(n)|^2}{E[|x(n)|^2]} \tag{4}$$

Where, E{.} is the Expectation Operator.

3.1. Motivation of Reducing PAPR

The troubles associated with OFDM, however, are also inherited by OFDMA. Hence, OFDMA also suffers from high peak to average power ratio (PAPR) because it is inherently made up of so many subcarriers [7]. The subcarriers are added constructively to form large peaks. High peak power requires High Power Amplifiers (HPA), A/D and D/A converters. Most radio systems employ the HPA in the transmitter to obtain sufficient transmission power. For the proposed of achieving the maximum output power efficiency, the HPA is usually operated at or near the saturation region [6].

Power efficiency is very necessary in wireless communication as it provides adequate area coverage, saves power consumption and allows small size terminals etc. It is therefore important to aim at a power efficient operation of the non-linear HPA with low back-off values and try to provide possible solutions to the interference problem brought about. Furthermore, the non-linear characteristic of the HPA is very responsive to the variation in signal amplitudes. The variation of OFDM signal amplitudes is very broad with high PAPR. Therefore, HPA will introduce inter-modulation between the different subcarriers and in- traduce additional interference into the systems due to high PAPR of OFDM signals. This additional interference leads to an increase in BER. In order to lessen the signal distortion and keep a low BER, it requires a linear work in its linear amplifier region with a large dynamic range. However, this linear amplifier has poor efficiency and is so expensive.

Therefore, a superior solution is to try to prevent the occurrence of such interference by reducing the PAPR of the transmitted signal with some manipulations of the OFDM signal itself [6]. Large PAPR also demands the DAC with enough dynamic range to accommodate the large peaks of the OFDM signals. Although, a high precision DAC supports high PAPR with a reasonable amount of quantization noise, but it might be very expensive for a given sampling rate of the system. Whereas, a low precision DAC would be cheaper, but its quantization noise will be significant, and as a result it reduces the signal Signal-to-Noise Ratio (SNR) when the dynamic range of DAC is increased to support high PAPR. In addition, OFDM signals show Gaussian distribution for large number of subcarriers, which means the peak signal quite rarely occur and uniform quantization by the ADC is not desirable. If clipped, it will introduce in-band distortion and out-of-band radiation (adjacent channel interference) in to the communication systems. Therefore, the best answer is to reduce the PAPR before OFDM signals are transmitted into nonlinear HPA and DAC.

4. CONVENTIONAL CLIPPING AND FILTERING

Amplitude Clipping and Filtering is one of the easiest techniques which may be under taken for PAPR reduction for an OFDM system. A threshold value of the amplitude is fixed in this case to limit the peak envelope of the input signal [8].

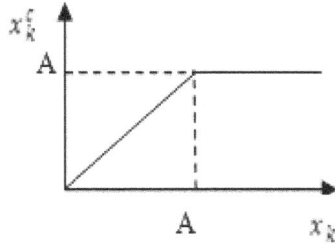

Figure 2. Clipping Function

The clipping ratio (CR) is defined as,

$$CR = \frac{A}{\sigma} \tag{5}$$

Where, A is the amplitude and σ is the root mean squared value of the unclipped OFDM signal. The clipping function is performed in digital time domain, before the D/A conversion and the process is described by the following expression,

$$x_k^c = \begin{cases} x_k & |x_k| \leq A \\ Ae^{j\phi(x_k)} & |x_k| > A \end{cases} \qquad 0 \leq k \leq N-1 \tag{6}$$

Where , x_k^c is the clipped signal, x_k is the transmitted signal, A is the amplitude and $\phi(x_k)$ is the phase of the transmitted signal x_k.

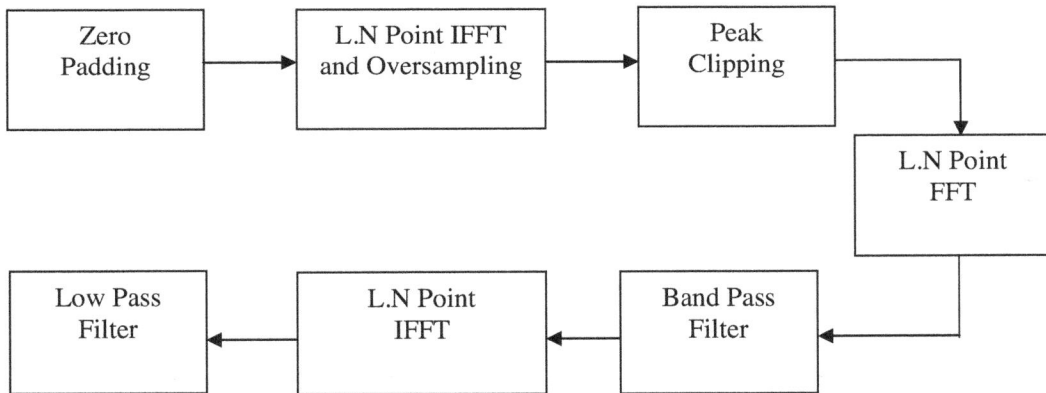

Figure 3. Block Diagram of PAPR reduction using Conventional Clipping & Filtering [9]

In the figure 3, the conventional block diagram of clipping & filtering method is shown which have some limitations described in below.

4.1. Limitations of Clipping and Filtering

➤ Clipping causes in-band signal distortion, resulting in BER performance degradation.

> ➤ Clipping also causes out-of-band radiation, which imposes out-of-band interference signals to adjacent channels. Although the out-of-band signals caused by clipping can be reduced by filtering, it may affect high-frequency components of in-band signal (aliasing) when the clipping is performed with the Nyquist sampling rate in the discrete-time domain[9].
> ➤ However, if clipping is performed for the sufficiently-oversampled OFDM signals (e.g., L ≥4) in the discrete-time domain before a low-pass filter (LPF) and the signal passes through a band-pass filter (BPF), the BER performance will be less degraded [10].
> ➤ Filtering the clipped signal can reduce out-of-band radiation at the cost of peak regrowth. The signal after filtering operation may exceed the clipping level specified for the clipping operation [3].

5. PROPOSED CLIPPING AND FILTERING SCHEME

Mentioning the third limitation that is clipped signal passed through the BPF causes less BER degradation, we design a new scheme for clipping & filtering method where clipped signal will pass through the high pass filter (HPF). This proposed scheme is shown in the figure 4. It shows a block diagram of a PAPR reduction scheme using clipping and filtering, where L is the oversampling factor and N is the number of subcarriers. The input of the IFFT block is the interpolated signal introducing $N(L-1)$ zeros (also, known as zero padding) in the middle of the original signal is expressed as,

$$X'[k] = \begin{cases} X[k], & for \;\; 0 \le k \le \frac{N}{2} \;\; and \;\; NL - \frac{N}{2} < k < NL \\ 0 & elsewhere \end{cases} \tag{7}$$

In this system, the L-times oversampled discrete-time signal is generated as,

$$x'[m] = \frac{1}{\sqrt{LN}} \sum_{k=0}^{LN-1} X'[k] e^{\frac{j2\pi m k}{LN}}, \qquad m = 0,1,...NL-1 \tag{8}$$

and is then modulated with carrier frequency f_c to yield a passband signal $x^p[m]$.

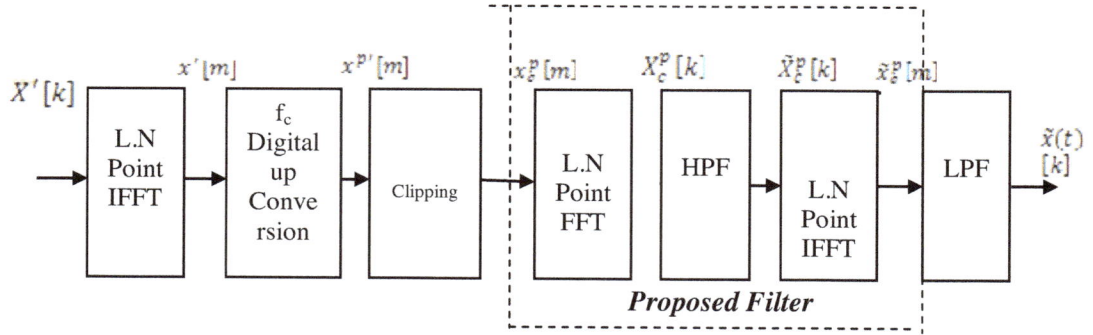

Figure 4. Block Diagram of Proposed Clipping & Filtering Scheme.

Let $x_c^p[m]$ denote the clipped version of $x^p[m]$, which is expressed as,

$$x_c^p[m] = \begin{cases} -A & x^p[m] \leq -A \\ x^p[m] & |x^p[m]| < A \\ A & x^p[m] \geq A \end{cases} \tag{9}$$

Or,

$$x_c^p[m] = \begin{cases} x^p[m] & if, \ |x^p[m]| < A \\ \dfrac{x^p[m]}{|x^p[m]|} \cdot A & otherwise \end{cases} \tag{10}$$

Where, A is the pre-specified clipping level. After clipping, the signals are passed through the proposed filter (Composed Filter). The filter itself consists on a set of FFT-IFFT operations where filtering takes place in frequency domain after the FFT function. The FFT function transforms the clipped signal $x_c^p[m]$ to frequency domain yielding $X_c^p[k]$. The information components of $X_c^p[k]$ are passed to a high pass filter (HPF) producing $\tilde{X}_c^p[k]$. This filtered signal is passed to the unchanged condition of IFFT block and the out-of-band radiation that fell in the zeros is set back to zero. The IFFT block of the filter transforms the signal to time domain and thus obtain $\tilde{x}_c^p[m]$.

6. DESIGN AND SIMULATION PARAMETERS

In this simulation, a linear-phase FIR filter using the Parks-McClellan algorithm is used in the composed filtering [4]. Existing method [6] uses the band pass filter. But, using this special type of high pass filter in the composed filter, significant improvement is observed in the case of PAPR reduction. The Parks-McClellan algorithm uses the Remez exchange algorithm and Chebyshev approximation theory to design filters with an optimal fit between the desired and actual frequency responses. The filters are optimal in the sense that the maximum error between the desired frequency response and the actual frequency response is minimized. It's a minimax in-band-zero filter which uses three independent equal errors. Table 1 shows the values of parameters used in the QPSK & QAM system for analyzing the performance of clipping and filtering technique [9]. We have simulated the both scheme in the same parameters at first. But, simulation results show the significant improvement occurs in PAPR reduction for proposed method.

Table 1. Parameters Used for Simulation of Clipping and Filtering.

Parameters	Value
Bandwidth (BW)	1 MHz
Over sampling factor (L)	8
Sampling frequency, f_s = BW*L	8 MHz
Carrier frequency, f_c	2 MHz
FFT Size / No. of Subscribers (N)	128
CP / GI size	32
Modulation Format	QPSK / QAM
Clipping Ratio (CR)	0.8, 1.0, 1.2, 1.4, 1.6

6.1. Simulation Results for PAPR Reduction

In this paper, simulation is performed for different CR values and compare with an existing method (Yong Soo et.al.) [9]. At first, we simulate the PAPR distribution for CR values =0.8, 1.0, 1.2, 1.4, 1.6 with QPSK modulation and N=128. Then, we simulate with QAM modulation and N=128 and compare different situations.

6.1.1 Simulation Results: (QPSK and N=128)

In the existing method, PAPR distribution for different CR value is shown in figure 5 (a). At CCDF $=10^{-1}$, the unclipped signal value is 13.51 dB and others values for different CR are tabulated in the table 2.

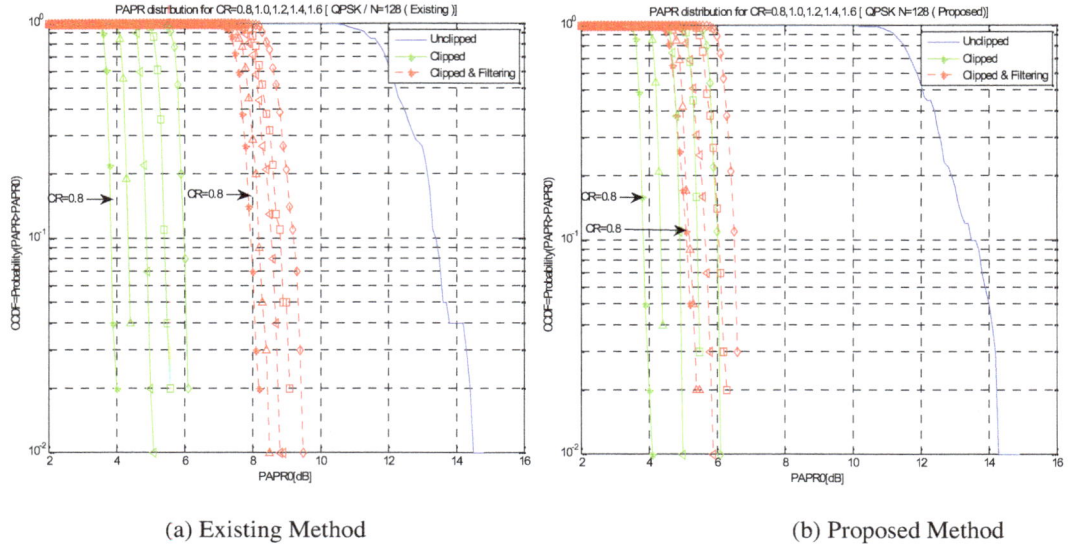

(a) Existing Method   (b) Proposed Method

Figure 5. PAPR Distribution for CR=0.8,1.0,1.2,1.4,1.6 [QPSK and N=128]

In the proposed method, simulation shows the significant reduction of PAPR is shown for different CR values in figure 5(b). At CCDF $=10^{-1}$, the unclipped signal value is 13.52 dB and others values for different CR are tabulated in the table 2. From table 2, we clearly observe that the proposed method reduces PAPR significantly with respect to existing Yong Soo [9] analysis. Proposed method of filter design is done with the same parameters that used in [9].

Table 2. Comparison of Existing with Proposed Method for PAPR value [QPSK and N=128]

CR value	PAPR value (dB) (Existing)	PAPR value (dB) (Proposed)	Improvement in PAPR value (dB)
0.8	7.94	5.11	2.83
1.0	8.19	5.18	3.01
1.2	8.55	5.65	2.90
1.4	8.81	6.04	2.77
1.6	9.22	6.51	2.71

6.1.2 Simulation Results: (QAM and N=128)

The simulation results are now shown for QAM modulation and no. of subscribers, N=128. In the existing method, PAPR distribution for different CR value is shown in figure 6 (a). At CCDF $=10^{-1}$, the unclipped signal value is 13.62 dB and others values for different CR are tabulated in the table 3. In the Proposed method, simulation shows the significant reduction of PAPR is shown for different CR values in figure 6(b). At CCDF $=10^{-1}$, the unclipped signal value is 13.64 dB and others values for different CR are tabulated in the Table 3.

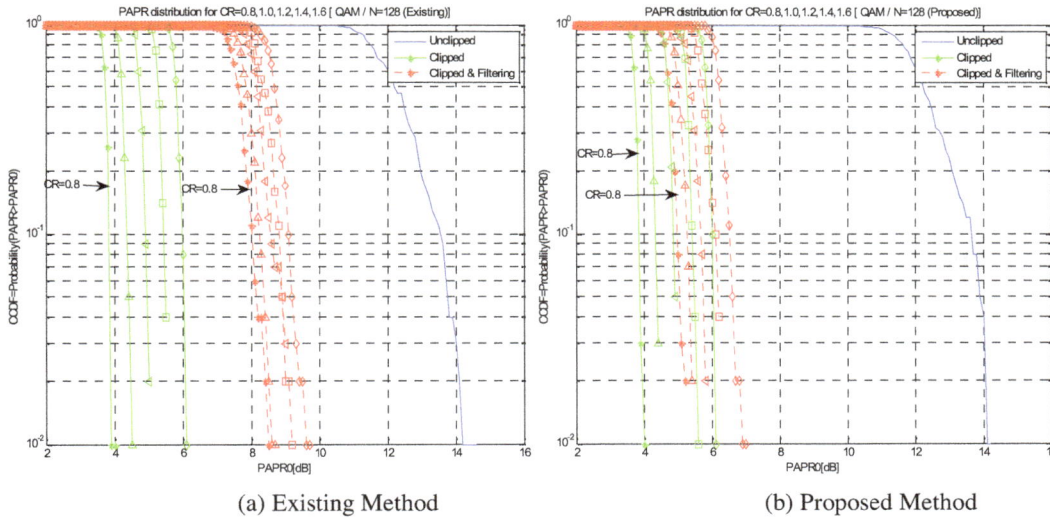

(a) Existing Method   (b) Proposed Method

Figure 6. PAPR Distribution for CR=0.8,1.0,1.2,1.4,1.6 [QAM and N=128]

The existing method and proposed method PAPR distribution values for different CR values are tabulated and differences are shown in table 3. From table 3, we clearly observe that the proposed method reduces PAPR significantly with respect to existing Yong Soo [9] analysis for QAM and N=128 also. So, proposed method works on efficiently for both QPSK & QAM.

Table 3. Comparison of Existing with Proposed Method for PAPR value [QAM and N=128]

CR value	PAPR value (dB) (Existing)	PAPR value (dB) (Proposed)	Improvement in PAPR value (dB)
0.8	8.08	4.97	3.11
1.0	8.24	5.25	2.99
1.2	8.56	5.67	2.89
1.4	8.81	6.09	2.72
1.6	9.10	6.51	2.59

Now, if we compare the values for different CR values in case of QPSK & QAM to show the effect of modulation on proposed filter design, it is observed that for the same number of subscribers (N=128) & low CR=0.8, QAM provides less PAPR than QPSK. But, at the moderate CR value (1.0, 1.2, 1.4), QPSK results less PAPR than QAM. At the high CR value=1.6, there is no differences between using QAM & QPSK. So, for low CR *(More Amount of Clipping)*, QAM is more suitable than QPSK in case of proposed filter.

6.2. Simulation Results for BER Performance

The clipped & filtered signal is passed through the AWGN channel and BER are measured for both existing & proposed methods. Figure 7 shows the BER performance when both clipping and clipped & filtering techniques are used. It can be seen from these figures that the BER performance becomes worse as the CR decreases. That means, for low value of CR, the BER is more.

6.2.1 Simulation Results: (QPSK and N=128)

At first, existing method is simulated for QPSK & N=128 for the same data mentioned in table 1 and resulted graph is shown in figure 7(a). Now, simulation is executed for proposed method using same parameters which is shown in figure 7(b) and observed that BER increases with respect to existing method for same value of CR.

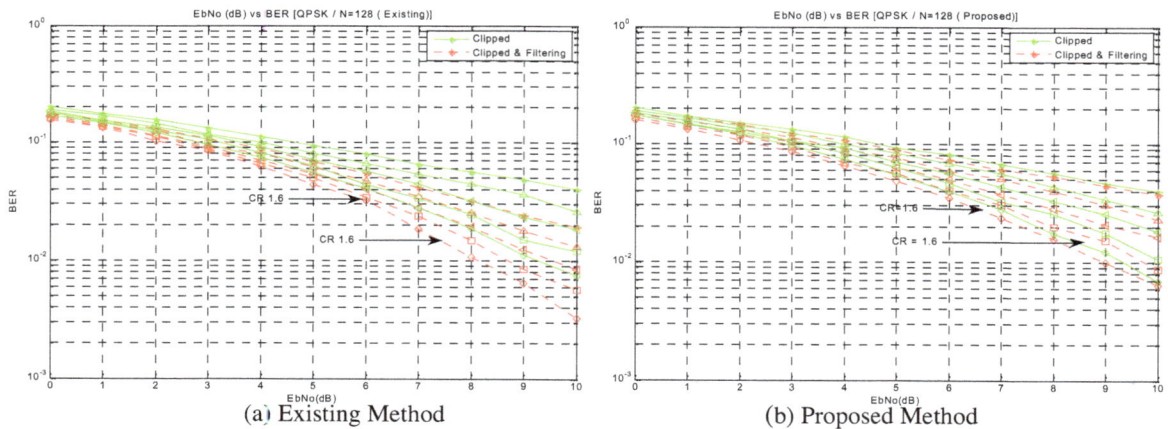

(a) Existing Method   (b) Proposed Method

Figure 7. BER Performance [QPSK and N=128]

From figure 7, it is observed that BER performance is worse in proposed method than existing method for different value of CR. The measured BER value at 6 dB point is tabulated in table 4.

Table 4. Comparison of BER value for Existing & Proposed Method [QPSK and N=128]

CR value	BER value (Existing)	BER Value (Proposed)	Difference in BER value
0.8	0.0528	0.0752	-0.0224
1.0	0.0451	0.0616	-0.0165
1.2	0.0396	0.0492	-0.0096
1.4	0.0344	0.0411	-0.0067
1.6	0.0289	0.0339	-0.0050

6.1.2 Simulation Results: (QAM and N=128)

At first, existing method is simulated for QAM & N=128 for the same data mentioned in table 1 and resulted graph is shown in figure 8(a). Now, simulation is executed for proposed method using same parameters which is shown in figure 8(b) and observed that BER increases with respect to existing method for same value of CR.

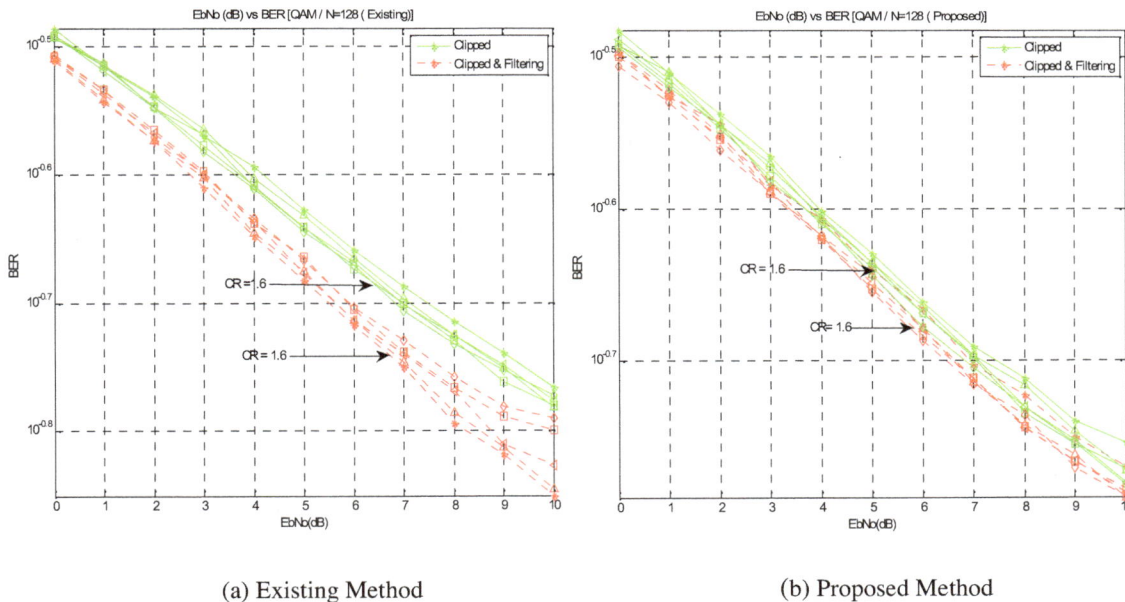

(a) Existing Method   (b) Proposed Method

Figure 8. BER Performance [QAM and N=128]

From figure 8, it is observed that BER performance is worse in proposed method than existing method for different value of CR. The measured BER value at 6 dB point is tabulated in table 5.

Table 5. Comparison of BER value for Existing & Proposed Method [QAM and N=128]

CR value	BER value (Existing)	BER Value (Proposed)	Difference in BER value
0.8	0.2004	0.2138	-0.0134
1.0	0.1994	0.2097	-0.0103
1.2	0.1951	0.2059	-0.0108
1.4	0.1923	0.208	-0.0157
1.6	0.1906	0.2049	-0.0143

BER performance is measured and compared in both the table 4 (QPSK) & table 5(QAM) which shows different CR value for both existing & proposed method in case of same parameter value. For large number of subscribers (N>128), it is found that existing method does not work properly. But, the proposed method works well. From table 4, it is observed that, for CR value (0.8,1.0,1.2,1.4 & 1.6) , the difference magnitude between existing & proposed method are 0.0224,0.0165,0.0096,0.0067 & 0.0050 respectively in QPSK. These BER degradations are acceptable as these are very low values. From table 5, it is observed that, for CR value (0.8,1.0,1.2,1.4 & 1.6) , the difference magnitude between existing & proposed method are 0.0134,0.0103,0.0108,0.0157 & 0.0143 respectively in QAM. These BER degradations are acceptable as these are very low values.

6. CONCLUSION

In this paper, a new scheme of amplitude clipping & filtering based PAPR reduction technique has been analyzed where PAPR reduces significantly compare to an existing method with slightly increase of BER. At first phase, simulation has been executed for existing method with QPSK modulation and number of subscriber (N=128) and then executed for the proposed method for same parameter and observed that PAPR reduces significantly. Next, the simulation has been performed for QAM modulation & N=128 and result shows the considerable improvement in case of QAM also. Then, the proposed method results for both QAM & QPSK modulation with N=128 has been compared. It is observed from the above comparison that for the same number of subscribers (N=128) & low CR=0.8, QAM provides less PAPR than QPSK. But, at the moderate CR value (1.0, 1.2, 1.4), QPSK results less PAPR than QAM. At the high CR value=1.6, there is no differences between using QAM & QPSK. So, for low CR *(More Amount of Clipping)*, QAM is more suitable than QPSK. In the present simulation study, ideal channel characteristics have been assumed. In order to evaluate the OFDM system performance in real world, multipath fading will be a consideration in future. The increase number of subscribers (N) & other parameters can be another assumption for further study.

REFERENCES

[1] Stefania Sesia, Issam Toufik and Matthew Baker, *LTE – The UMTS Long Term Evolution: From Theory to Practice",* United Kingdom : John Wiley & Sons Limited,2009.

[2] R. Prasad, *OFDM for Wireless Communications Systems.* London, Boston: Artech House, Inc, 2004.

[3] S. H. Han and J. H. Lee, "An overview of peak-to-average power ratio reduction techniques for multicarrier transmission", *IEEE Wireless Comm,* vol. 12, no.2, pp.56-65, Apr. 2005.

[4]   M. Mowla, " Peak to average power ratio analysis and simulation in LTE system", *M.Sc dissertation,* Department of Electrical &   Electronic Engineering, Rajshahi University of Engineering & Technology, Bangladesh, May 2013.

[5]   Muhammad Atif Gulzar , Rashid Nawaz and Devendra Thapa, "Implementation of MIMO-OFDM System for WiMAX*", M.Sc Dissertation*, Department of Electrical Engineering, Linnaeus University, Sweden, 2011.

[6]   T. Jiang and Y. Wu, "An Overview: peak-to-average power ratio reduction techniques for OFDM signals", *IEEE Trans. Broadcast.,* vol. 54, no. 2, pp. 257-268, June 2008.

[7]   G.Sikri and Rajni, "A Comparison of Different PAPR Reduction Techniques in OFDM Using Various Modulations", *International Journal of Mobile Network Communications & Telematics (IJMNCT)* vol.2, no.4, pp. 53-60, 2012.

[8]   Natalia Revuelto,"PAPR reduction in OFDM systems", *M.Sc Dissertation,* Universitat Politecnica de Catalunya, Spain, 2008.

[9]   Y.S. Cho, J. Kim, W.Y.Yang and C.G. Kang, *MIMO OFDM Wireless Communications with MATLAB*, Singapore: John Wiley & Sons (Asia) Pte Ltd, 2010.

[10] H.Ochiai and K.Imai, "On clipping for peak power reduction of OFDM signals". *IEEE GTC* , vol.2, pp.731–735,2000.

[11] S.S.Anwar, S.L.Kotgire, S.B.Deosarker and D.E. Rani, " New Improved Clipping and filtering technique algorithm for peak-to-average power ratio reduction of OFDM signals", *International Journal of Computer Science and Communication,* vol.3, no. 1,pp. 175-179, January-June 2012.

3

REAL-TIME MODE HOPPING OF BLOCK CIPHER ALGORITHMS FOR MOBILE STREAMING

Kuo-Tsang Huang[1], Yi-Nan Lin[2] and Jung-Hui Chiu[1]

[1]Department of Electrical Engineering, Chang Gung University, Tao-Yuan, Taiwan
[2]Department of Electronic Engineering, Ming Chi University of Technology, New Taipei City, Taiwan
`d9221006@gmail.com, jnlin@mail.mcut.edu.tw, jhchiu@mail.cgu.edu.tw`

ABSTRACT

It has been shown that the encrypted information or ciphertext produced by symmetric-key block ciphers with Electronic codebook mode is vulnerable to ciphertext searching, replay, insertion and deletion because it encrypts each block independently. To compensate for this, each block of the encrypted information should be encrypted dependently. The encrypted information should be operated with a special mode. The operation mode should be changed. This paper analysis what an operational mode of block ciphers needs to feedback exactly and proposes a simple real-time changing operation mode technique that extends the existing mode changing opportunity. The new change operation mode technique considers the sign differences between the intra-feedback information and the public-feedback information, and then adaptively determines the corresponding change operation mode factor for each data block. This mode hopping technique for mobile streaming security is highly suitable for recent block computing in future various environments.

KEYWORDS

Mobile, Stream, Block Cipher, Mode of Operation, Feedback, Hopping

1. INTRODUCTION

Habits of personal records in the past were drawing and making perfect life memory in order to cross over the distance of time or space. Personal records of life memory can sample and collect to retain constantly and be treasured in future moment taste. The intersection of social community and personal information sharing development of stress instantaneity and emphasis on real-time acknowledge, e.g. a GOOD respond. Digital camera surrounding analogy affairs complete digital record, via the mobile communication networks across space restrictions immediately to share personal things with specific groups or the general public.

Security has become an important issue on embedded systems due to the popularity. The typical security requirements across a wide range of embedded systems are user identification, secure network access, secure communications, secure storages, content security and availability. Modern applied cryptography in the communications networks demands high processing rate to fully utilize the available network bandwidth.

Block encryption may be vulnerable to ciphertext searching, replay, insertion, and deletion because it encrypts each block independently. Unfortunately, the non-feedback conventional block ciphers have plaintext-ciphertext pair problem with the disadvantage of limit block region scramble. A disadvantage of this method is that identical plaintext blocks are encrypted into identical ciphertext blocks. Non-feedback Conventional block ciphers do not hide data patterns well. A striking example of the degree to which non-feedback electronic codebook (ECB) mode can leave plaintext data patterns in the ciphertext can be seen when the electronic codebook mode is used to encrypt a bitmap image which uses large areas of uniform colour. Block cipher modes of encryption beside the electronic codebook mode have been suggested to remedy these drawbacks.

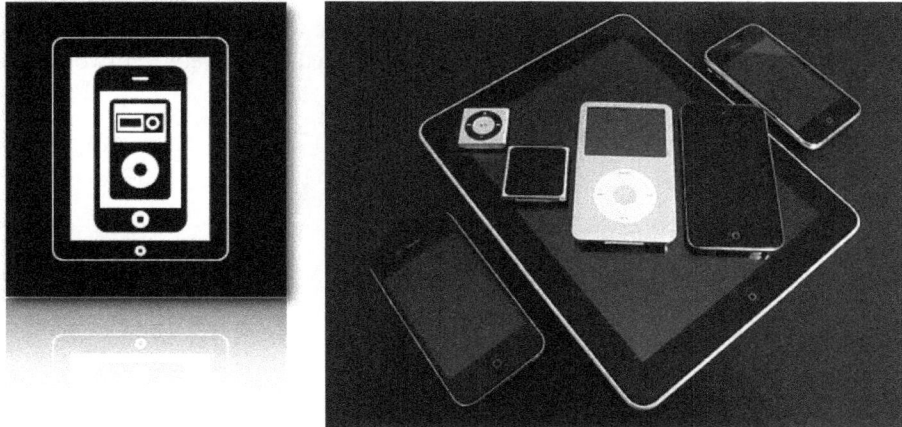

Figure 1. Everywhere Smart Equipments such as iDevice [1]

Modes of operation are the procedure of enabling the repeated use of a block cipher under a single key. Block cipher modes of encryption beside Electronic codebook (ECB) have been suggested to remedy these drawbacks. In Internet, modern protocols support several operation modes and ciphers to provide the varied choice for suitable various operated environments. For example Secure Socket Layer (SSL) [2] and Internet Protocol Security (IPSec) [3] support multi-cipher and multi-mode [4][5][6][7][8].

The standard modes of operation described in the literature [9][10] provide confidentiality. How to choose an appropriate operation mode? The different mode has the different characters. For example, both of CFB and OFB can be design operating without padding with bit-based size keystream output; both of CBC and CFB can self sync to avoid channel noise error propagation; and both of CFB and OFB encryption and decryption applications need an encryption module only to reach both usages. In addition, only the forward cipher function of the block cipher algorithm is used in both encryption and decryption operations, without the need for the inverse cipher function. However, one disadvantage still exists with these techniques when they are applied to practice encryption: any changing mode in encrypting session and between encrypting blocks requires the re-configuration of crypto module and has no mechanism guaranteeing a smooth transition in such encrypting single session.

Our idea of the improvement security of mobile streaming is making mode of operation hopping like frequency copping in communication. In this paper, a new technique, called feedback exactly technique, is proposed. The technique uses possible options to satisfy any next one of all feedbacks so that can real-time change from any one mode to another, even during the same session. The rest of this paper is organized as follows. Section 2 contains acknowledge of block cipher, multi-mode crypto ASIC, and mobile streaming protections. Section 3 describes an application which applies to dynamically exchange mode of operation. Hopping mode of operation improves mobile streaming security in communication. The idea is based on the proposed technique of feedback exactly as what an operational mode of block ciphers needs, described in the Section 4. Section 5 contains designs of the smooth changing process and seamless changeable system architecture of crypto module. Finally Section 6 is the conclusions.

2. RELATED WORKS

Symmetric-key ciphers use trivially related cryptographic secret keys for both encryption of plaintext and decryption of ciphertext. The secret encryption key is trivially related to the secret decryption key, in that they may be identical or there is a simple transformation to go between the two secret keys. In practice, the secret keys represent a shared secret between two or more parties that can be used to maintain a private information link.

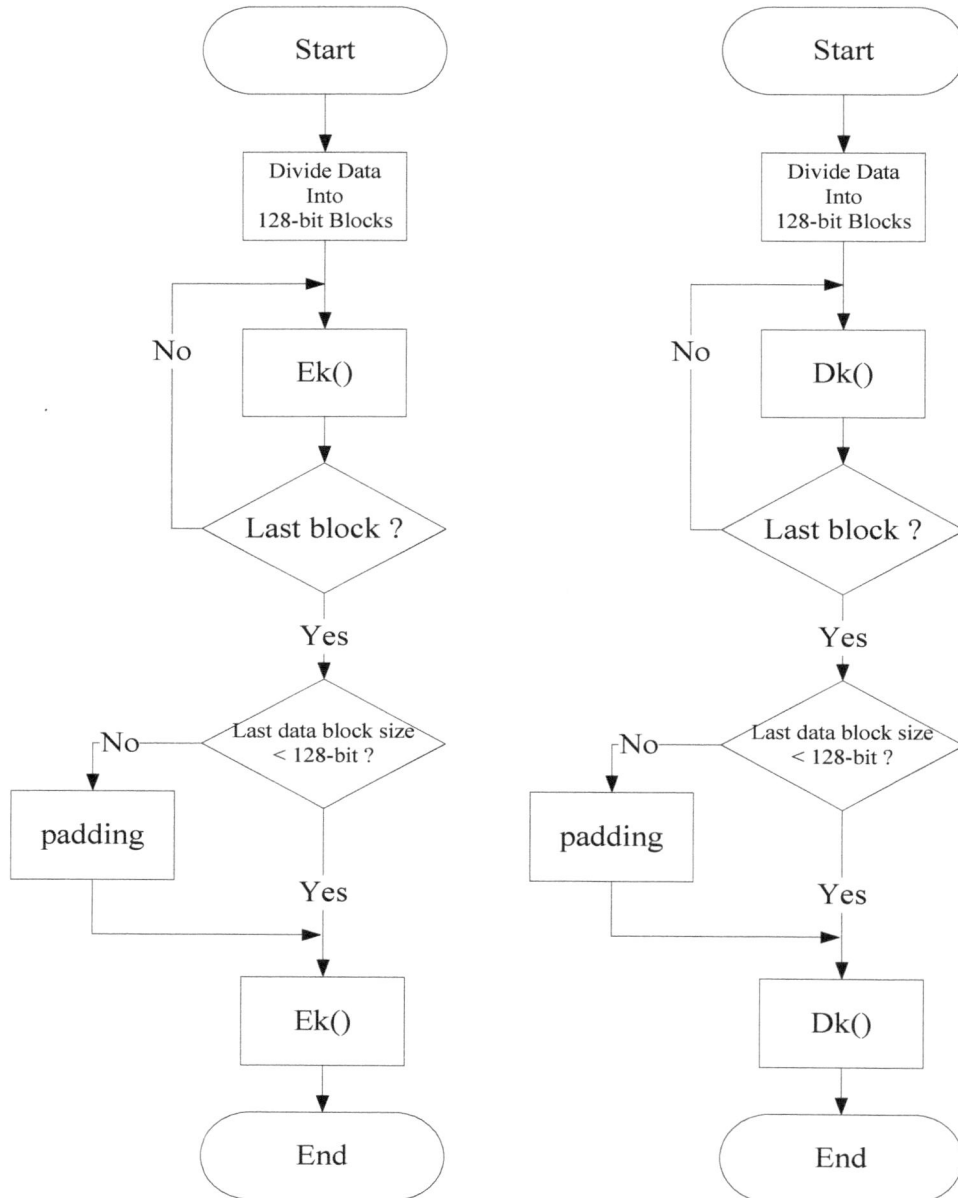

Figure 2. Flow Diagram of Encryption and Decryption for Block Cipher

2.1. Block Ciphers and Padding

The major concept in information security today is to continue to improve encryption algorithms. There are two major types of encryption algorithms for cryptography: symmetric-key algorithms and public-key algorithms. Symmetric-key algorithms also referred to as conventional encryption algorithms or single-key encryption algorithms are a class of algorithms that use the same cryptographic keys for both encryption of plaintext and decryption of ciphertext. It remains by far the most widely used of the two types of encryption algorithms. Symmetric-key encryption algorithms can use either stream ciphers or block ciphers. Block ciphers take a number of bits and encrypt them as a single unit, padding the plaintext so that it is a multiple of the block size. Blocks of 64 bits have been commonly used. The Advanced Encryption Standard (AES) algorithm [14] approved by NIST in December 2001 uses 128-bit blocks. In the normal separation of encryption and decryption, the author separate encryption and decryption processing when receives messages from input port dependent on the target task.

A block cipher works on units of a fixed block size but messages come in a variety of lengths. So that the final block be padded before encryption. Several padding schemes exist. The simplest is to add null bytes to the plaintext to bring its length up to a multiple of the block size, but care must be taken that the original length of the plaintext can be recovered; this is so, for example, if the plaintext is a C style string which contains no null bytes except at the end. Figure 2 is the flow diagram of encryption and decryption for block cipher with padding when last block is not full of bit.

2.2. Mobile Streaming Protections

Business practices of digital content distribution network are most of the structure of Digital Rights Management (DRM) protected [20]. DRM mechanism can protect digital contents against unauthorized user access. From the aspect of the video streaming contents, the encryption mechanism is the protective skill mainly; meanwhile, the encryption methods vary due to the form of video streaming contents. Users receive encrypted digital content through various licensing relationships, which called Rights Object (RO), to distinguish different access control rights. Encryption mechanisms in addition to confidentiality protection in the transmission process, if any change in information will cause decryption failed and, therefore, be able to tell whether a message has been tampered with. Encryption and decryption for protection of information simplify key management issues.

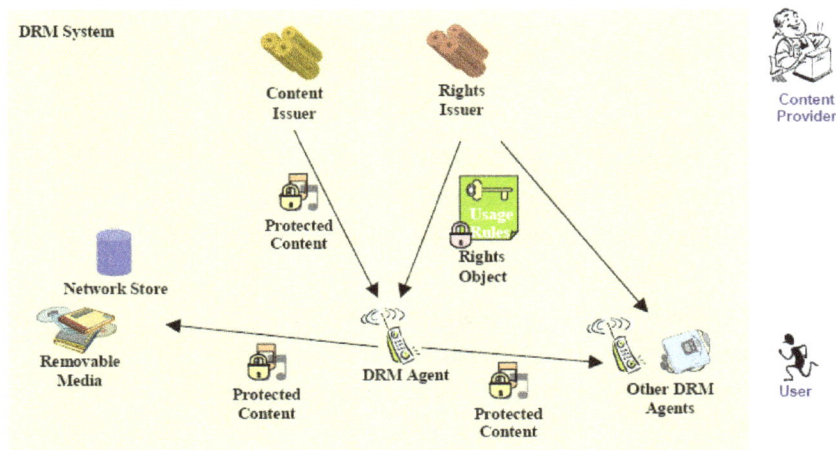

Figure 3. Digital Rights Management Architecture of Open Mobile Alliance [20]

Multimedia security is often used software encryption to ensure the safety of the computing complexity and provide the security requirements of confidentiality. However, the actual use of the classic block cipher (such as AES [14]) to encrypt the entire video stream requires a lot of computing time, and many multimedia applications (such as TV broadcast) require only a low level of security. People need to develop specifically for this Fast encryption target environment.

Transmission of multimedia services with security, such as: secure transmission of visual applications (video clients, video conferencing and other mobile devices ...), variant clients may experience, or low-power devices in the sending or receiving end, such as mobile devices or handheld devices, will encounter problems plagued low-cost devices of receivers [21]. In today's practice, there are two possible solutions: one is a low computational complexity of encryption and decryption algorithms to replace [22], but this will reduce the privacy concerns; the other one is the selective encryption [23][24], encryption is only part of that section of the data, hope to achieve while maintaining safety results. Selective encryption seems to be only part of the contents to ensure the confidentiality, explicitly part of the contents dispersed in the environment. But fortunately the characteristics of video streaming viewing audience must have a smooth flow of quality assurance, an intermittent stream video multimedia content, is contrary to users feeling pain. However, some choose to encrypt which target parts? Alattar etc. [24] proposed a multimedia digital content encryption algorithm for low energy and computing powers handheld devices. For security analysis, they proposed two ciphertext attacks discussed and introduced the technology environment and conditions apply. In the field of multimedia security, encryption for the entire video stream using classical ciphers (such as AES [14]) requires a lot of computing time. The selective encryption algorithms for part of the contents accelerate the existing computing programs. These methods do not provide maximum security but only in the computational complexity of security and balance.

3. THE IMPROVEMENT OF MOBILE STREAMING SECURITY

Cipher modes of operation extend the limit region and scrambling to the post-plaintext message. The formulas of five conventional modes, non-feedback electronic codebook (ECB) mode, cipher block chaining (CBC) mode, output feedback (OFB) mode, cipher feedback (CFB) mode, Counter (CTR) mode, of block cipher for confidentiality could be expressed with the following notations and ciphering Table 1.

Notations:

Ek( )   the Encryption algorithm of the block cipher with session key K

Dk( )   the Decryption algorithm of the block cipher with session key K

Pi   the i-th block of Plaintext with respect to the operation mode

Ci   the i-th block of Ciphertext with respect to the operation mode

Yi   the i-th output State (or keyStream) of the mode like OFB

CTRi   the i-th CounTeR value for the CTR mode of the block cipher

IV   the Initial Value needed for the different operation mode

Table 1. (a) Enciphering for the Conventional Block Cipher Mode of Operation

mode of operation	ciphertext formula	extra input
ECB	$C_i = E_k(P_i)$	(not necessary)
CBC	$C_i = E_k(P_i \oplus C_{i-1})$	C_{i-1} with $C_0 = IV_1$
CFB	$C_i = P_i \oplus E_k(C_{i-1})$	C_{i-1} with $C_0 = IV_1$
OFB	$C_i = P_i \oplus E_k(Y_{i-1})$	Y_{i-1} with $Y_0 = IV_2$
CTR	$C_i = P_i \oplus E_k(CTR_{i-1})$	CTR_i with $CTR_0 = IV_3$

with $C_0 = IV_1$, $Y_0 = IV_2$, $CTR_0 = IV_3$, and $CTR_i = IV_3 + i$

Table 1. (b) Deciphering for the Conventional Block Cipher Mode of Operation

mode of operation	plaintext formula	extra input
ECB	$P_i = D_k(C_i)$	(not necessary)
CBC	$P_i = D_k(C_i) \oplus C_{i-1}$	C_{i-1} with $C_0 = IV_1$
CFB	$P_i = C_i \oplus E_k(C_{i-1})$	C_{i-1} with $C_0 = IV_1$
OFB	$P_i = C_i \oplus E_k(Y_{i-1})$	Y_{i-1} with $Y_0 = IV_2$
CTR	$P_i = C_i \oplus E_k(CTR_{i-1})$	CTR_i with $CTR_0 = IV_3$

with $C_0 = IV_1$, $Y_0 = IV_2$, $CTR_0 = IV_3$, and $CTR_i = IV_3 + i$

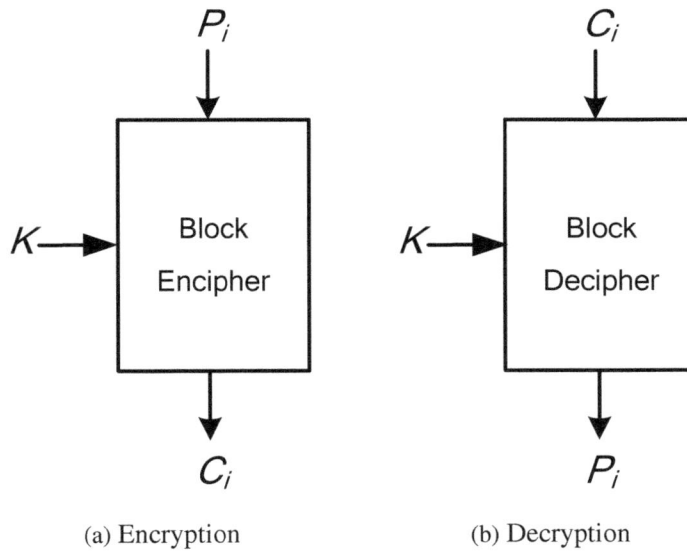

(a) Encryption   (b) Decryption

Figure 4. Conventional Block Ciphering

The block diagrams of the enciphering and deciphering formulas are shown in Figures 5(a) - (e).

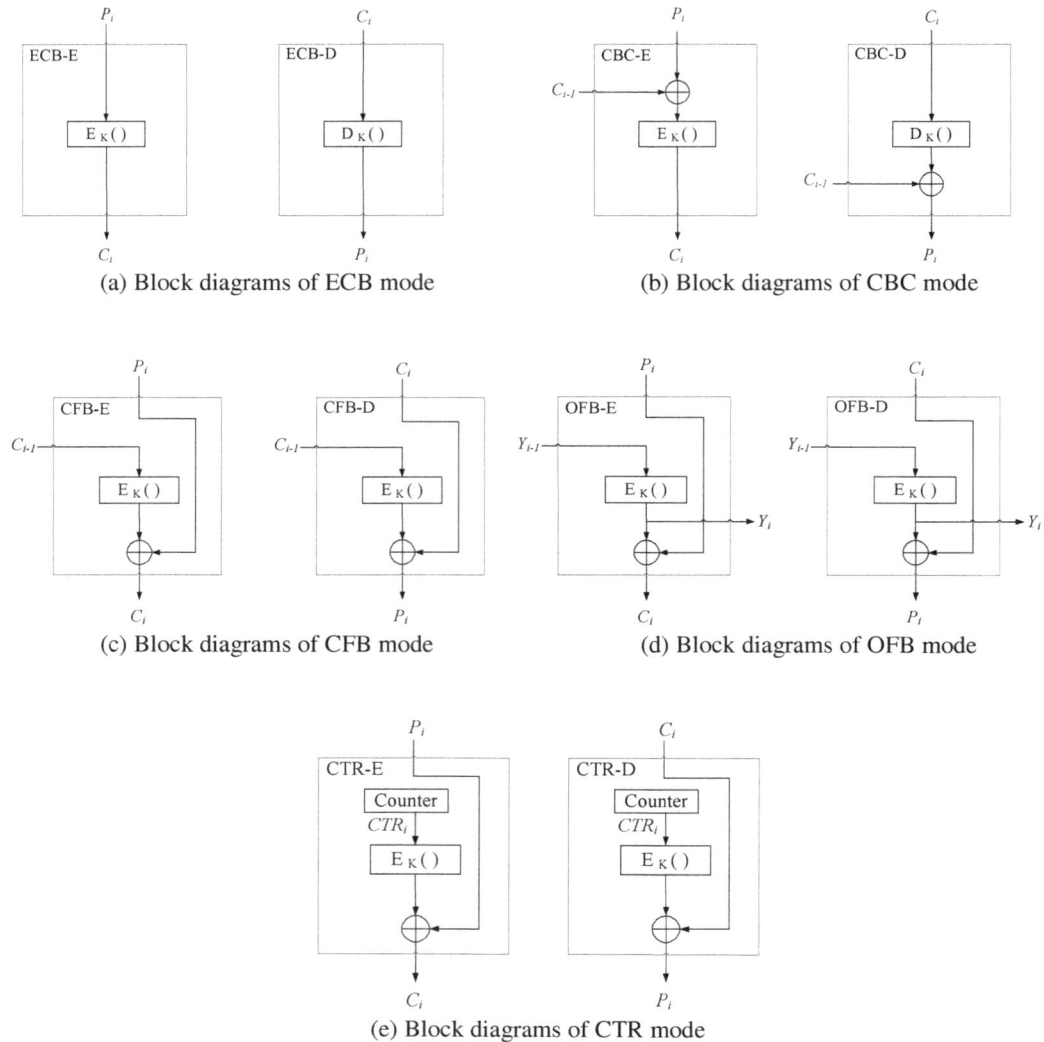

(a) Block diagrams of ECB mode

(b) Block diagrams of CBC mode

(c) Block diagrams of CFB mode

(d) Block diagrams of OFB mode

(e) Block diagrams of CTR mode

Figure 5. Block Diagrams of Different Ciphering Modes

3.1 Problem Descriptions

According the above expressions, except the ECB mode, encipher needs the regular input of plaintext Pi and the extra input (C_{i-1}, Y_{i-1}, or CTR_{i-1}) to perform the ciphering formula to find the output ciphertext C_i. The extra input is not the same for all the modes of operation. This situation will make it not suitable for changing the mode of operation during the ciphering process. In other words, the direct change immediately from one mode to another in a session is impossible. For instance, if the current mode is ECB, it can't change to the OFB mode, because that it does not have the extra input Y_{i-1} which is necessary for the OFB mode to perform correctly.

Based on the conventional ciphertext/plaintext formula, it could be configured to one of the five modes of operation, and performs the ciphering task during the entire session with this fixed mode. After that, it could be reconfigured to another mode of operation for the requirement of the subsequent job. In this section, the author introduce encrypting bursts with different ciphers or different operation modes. One example of the problem is shown in following.

3.2 An Example of Internal Data Feedback Problem

If users want to dynamically change between blocks in a session of encrypt tasks, they must prepare all the feedbacks to suit one of the next block needed. For example, a user must prepare the current ciphertext data block feedbacks, if next block would operate the CBC mode. In current block slot, prepare several possible options for the next block operation needed is a direct solution and can solve the most of feedback problems, but not the special case of internal data feedback problem. Feedback problems include one of the special case the author called internal data feedback problem. The most famous existing example is output feedback problem (i.e., a previous slot block data feedback to the same one single cipher module circularly). The index i is timing slot index in Figure 6. The slot unit of data is block length size.

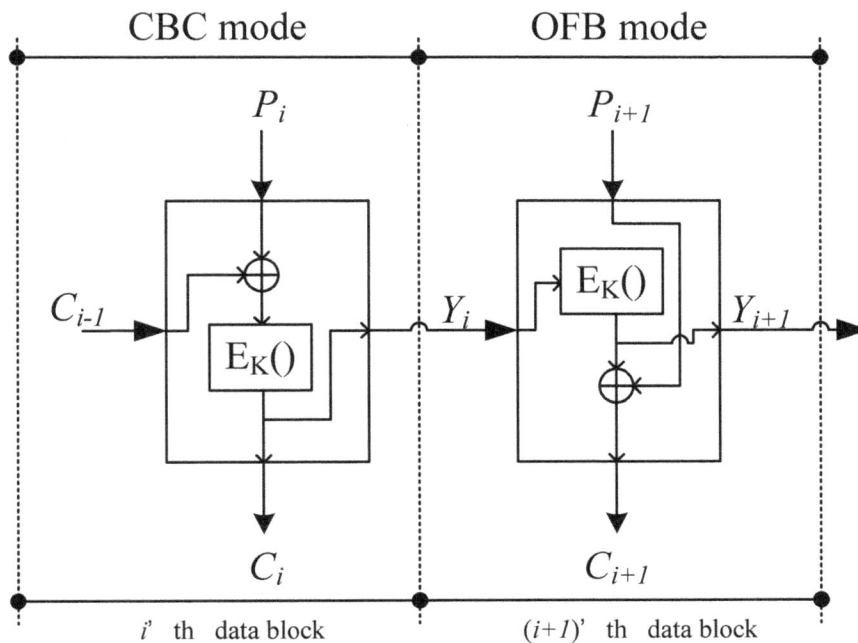

$$C_i = E_K[P_i \oplus C_{i-1}] \qquad C_{i+1} = E_K[Y_i] \oplus P_{i+1}$$

(a) In encrypting, mode of block operation changing from CBC to OFB

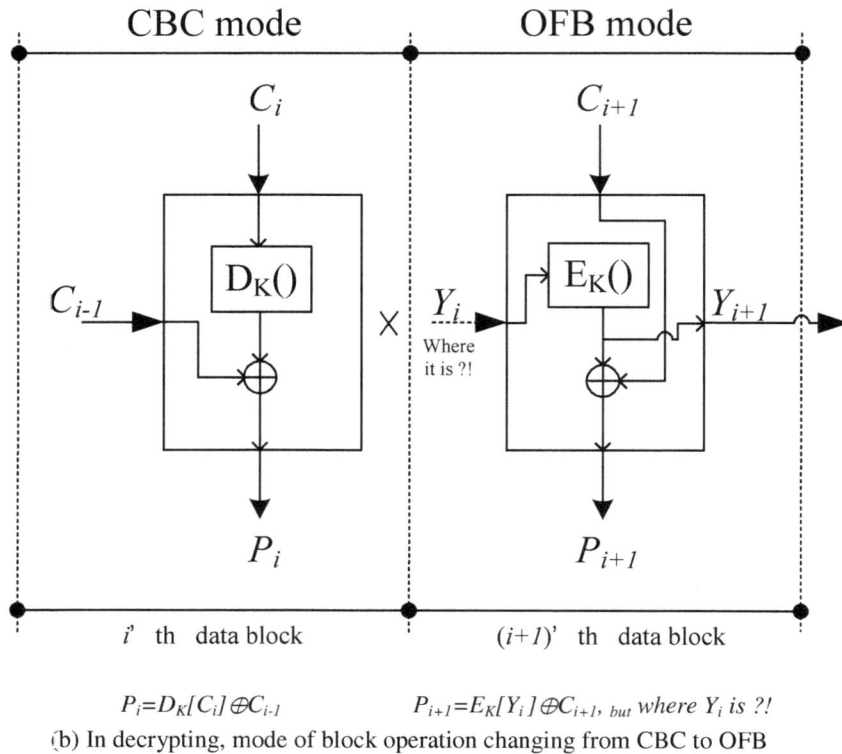

$$P_i=D_K[C_i]\oplus C_{i-1}\qquad\qquad P_{i+1}=E_K[Y_i]\oplus C_{i+1},\ _{but}\ where\ Y_i\ is\ ?!$$

(b) In decrypting, mode of block operation changing from CBC to OFB

Figure 6. An Example of Internal Data Feedback Problem

4. FEEDBACK EXACTLY AS WHAT AN OPERATIONAL MODE OF BLOCK CIPHERS NEEDS

Modern block cipher protocols support five modes of operation, ECB, CBC, OFB, CFB, CTR, to provide the confidentiality for the requirements of different applications. Since each mode has its special plaintext/ciphertext formula with different extra input, it could not change the mode of operation anywhere at any time.

By inspecting formulas and extra inputs in Table 1 and block diagrams in Figure 5, it could be found out that ECB, CBC, and CFB modes have lack of stream variable Si. The assignment of the stream variable Si for ECB, CBC, and CFB modes is necessary. To solve the above inconsistent extra input problem, the union the extra inputs of different modes of operation is set to be the new extra inputs, and be redefined as the previous status $<P_{i-1}, C_{i-1}, Y_{i-1}>$. The status is defined as the 3-tuples vector of the form <plaintext Pi, ciphertext Ci, and internal state Y_i> = $<P_i, C_i, Y_i>$. After finishing the ciphering operation, the previous status $<P_{i-1}, C_{i-1}, Y_{i-1}>$ is updated and replaced the current status $<P_i, C_i, Y_i>$. In this sense, the conventional enciphering formula could be considered as the ciphertext updating formula, and the conventional deciphering formula could be considered as the plaintext updating formula. In order to complete the updating procedure, the update of the **internal feedback exactly state S_i** is introduced here for the different modes of operation. The internal feedback exactly state Si could be found by the general function $g_{mc}(\ )$ of the mode of operation specified by the mode code m_c.

$$S_i= g_{mc}(k, P_{i-1}, C_{i-1}, Y_{i-1}, CTR_{i-1}) \ ...(1)$$

For the ECB or CBC mode, it does not have the internal feedback exactly state S_i in the conventional form. Hence, the internal feedback exactly state S_i should be defined as to be any variable got from the ECB structure, such as the ciphertext, or that of the next-to-last round. For simplicity, let $S_i = S_{i-1}$.

For the CFB, OFB and CTR modes, they are the stream cipher like. The internal feedback exactly state S_i could be defined as to be the keystream variable. Hence, the ciphertext would be expressed as $C_i = P_i \oplus S_i$, and the internal feedback exactly state S_i could be expressed as $S_i = E_k (C_{i-1})$, $S_i = E_k (S_{i-1})$, $S_i = E_k (CTR_{i-1})$ for CFB, OFB, CTR, respectively.

Any assignment of S_i for ECB, CBC, and CFB modes could be workable as long as S_i are kept the same at encipher and decipher. For simplicity, the assignments of S_i would be $S_i = S_{i-1}$ for ECB and CBC modes, and $S_i = E_k(C_{i-1})$ for CFB mode, as shown in Table 2.

Enciphering:

$f_{mc}( )$ is the enciphering function of the mode of operation specified by the mode code m_c.

$$C_i = f_{mc}(k, P_i, P_{i-1}, C_{i-1}, S_{i-1}) \quad ...(2)$$

Table 2. (a) Enciphering for mode hopping of the improved block cipher mode of operation

operation mode	mode code m_c	enciphering formula (status updating formula)	previous status
ECB	000	$C_i = E_k(P_i)$, $S_i = S_{i-1}$	$P_{i-1}, C_{i-1}, S_{i-1}$
CBC	001	$C_i = E_k(P_i \oplus C_{i-1})$, $S_i = S_{i-1}$	$P_{i-1}, C_{i-1}, S_{i-1}$
CFB	010	$C_i = P_i \oplus S_i$, $S_i = E_k (C_{i-1})$	$P_{i-1}, C_{i-1}, S_{i-1}$
OFB	011	$C_i = P_i \oplus S_i$, $S_i = E_k (S_{i-1})$	$P_{i-1}, C_{i-1}, S_{i-1}$
CTR	100	$C_i = P_i \oplus S_i$, $S_i = E_k (CTR_{i-1})$	$P_{i-1}, C_{i-1}, S_{i-1}$

with $C_0 = IV_1$, $S_0 = IV_2$, $CTR_0 = IV_3$, and $CTR_i = IV_3 + i$

Deciphering :

$f_{mc}^{-1}( )$ is the deciphering function of mode m_c, and is the inverse function of $f_{mc}( )$.

$$P_i = f_{mc}^{-1}(k, C_i, P_{i-1}, C_{i-1}, S_{i-1}) \quad ...(3)$$

Table 2. (b) Deciphering for mode hopping of the improved block cipher mode of operation

operation mode	mode code m_c	deciphering formula (status updating formula)	previous status
ECB	000	$P_i = D_k(C_i)$, $S_i = S_{i-1}$	$P_{i-1}, C_{i-1}, S_{i-1}$
CBC	001	$P_i = D_k(C_i) \oplus C_{i-1}$, $S_i = S_{i-1}$	$P_{i-1}, C_{i-1}, S_{i-1}$
CFB	010	$P_i = C_i \oplus S_i$, $S_i = E_k (C_{i-1})$	$P_{i-1}, C_{i-1}, S_{i-1}$
OFB	011	$P_i = C_i \oplus S_i$, $S_i = E_k (S_{i-1})$	$P_{i-1}, C_{i-1}, S_{i-1}$
CTR	100	$P_i = C_i \oplus S_i$, $S_i = E_k (CTR_{i-1})$	$P_{i-1}, C_{i-1}, S_{i-1}$

with $C_0 = IV_1$, $S_0 = IV_2$, $CTR_0 = IV_3$, and $CTR_i = IV_3 + i$

The block diagrams of the improved enciphering and deciphering formulas are shown in Figures. 7(a) - (e).

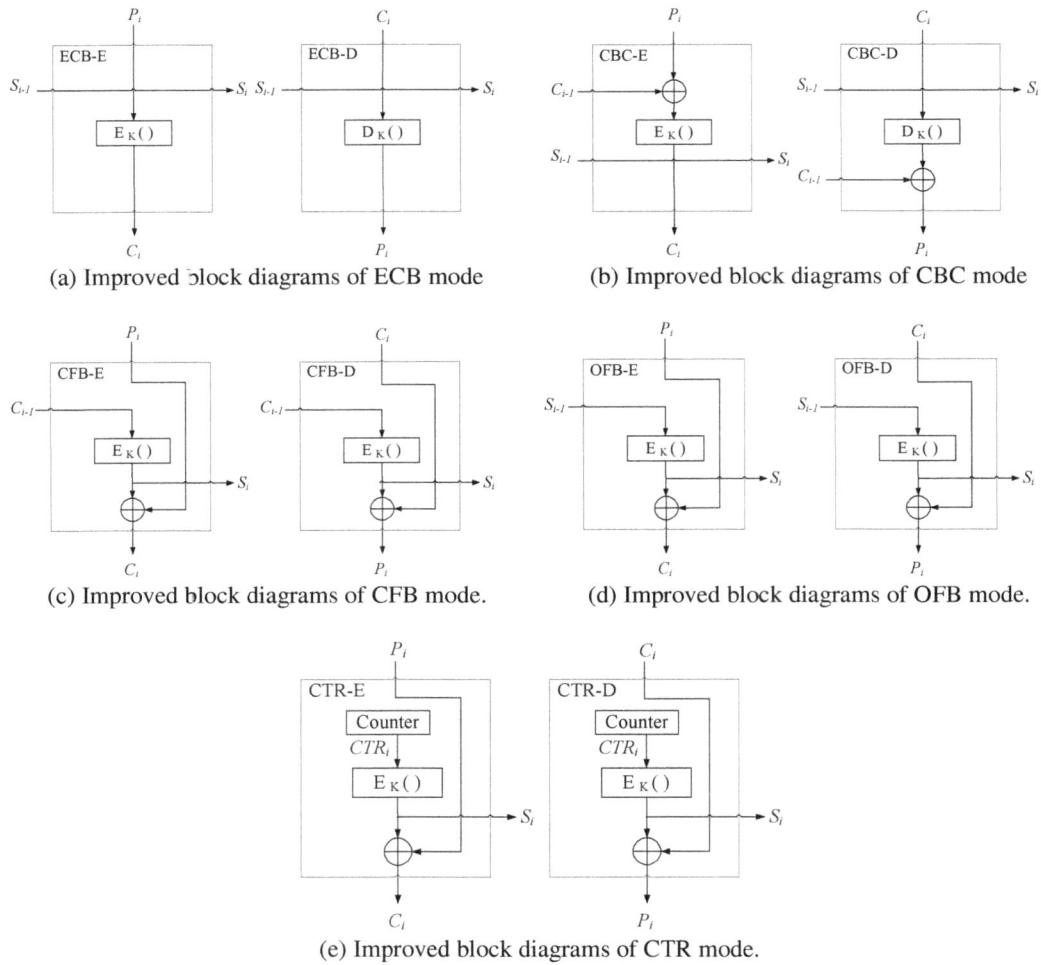

(a) Improved block diagrams of ECB mode

(b) Improved block diagrams of CBC mode

(c) Improved block diagrams of CFB mode.

(d) Improved block diagrams of OFB mode.

(e) Improved block diagrams of CTR mode.

Figure 7. Improved Block Diagrams of Different Ciphering Modes

5. APPLY TO REAL-TIME CHANGING OPERATIONAL MODE

Some symmetric ciphers use the same function to perform encryption and decryption applications, some does not. Because of some modes decrypt message using the same algorithm as encrypting, i.e. CFB and OFB. The author defines three levels about module, function, and application to distinguish different usage situations. A crypto cipher module has two functions, an encryption function and a decryption function. E-application is a practical one in message encryption, and D-application is a practical one in message decryption. E-function is a mapping function of transforming information (referred to as plaintext) to make it unreadable to anyone except those possessing special knowledge, usually referred to as a key. D-function is the inverse one of E-function. Table 3 indicates the notations what the author need.

When this simple real-time changing operational mode technique is used for message encryption, the mode change depends on a control sequence. The target sequence is synchronous between encryption and decryption parts and it is pre-shared as a secret. One can generate many different control sequence configuration files at design time for peers which can have a great diversity to change the modules. Before a session of communication begins, the communication parties need to negotiate a control sequence configuration file introduced by adaptive secure protocols such as IPSec. However, the additional cost of a target control sequence can be flexibly reduced according to the practical requirements of security.

If user wants guarantee real-time changing mode of crypto operations, a technique must allow keep one step ahead. The author use AES (Daemen et al, 2002) to illustrate exchange procedures. A secret key K with a 128 bit length is required for the encryption and decryption (denoted by E_K and D_K respectively) of AES (Daemen et al, 2002). Suppose the author want to encrypt a long message sequence, P^T. First, the author divide the sequence into several blocks, P_i, $i=1,2...$, each P_i with 128 bit length. Then, the encryption operations are performed by Figure 8(a) and the decryption operations are performed by Figure 8(b). Where S_i, for $i=1,2...$, is a 128 bit sequence.

Both enciphering and deciphering algorithms are needed implemented within the deciphering improved structure, but only one of them is used for each mode of operation. To save the consuming power, in the deciphering improved structure, the enciphering box is disabled for ECB, CBC modes and enabled for CFB, OFB, CTR modes; and similarly, the deciphering box is enabled for ECB, CBC modes and disabled for CFB, OFB, CTR modes. The author can get that it is flexible, workable, and applicable in modern communication applications.

Table 3. Notations

B	A crypto **module** of any one symmetric cipher which can perform both the encryption and decryption **applications**. This module has two **functions**, an encryption function and a decryption function.
B_i	Use B in an encryption application (E-application) at a i'th block slot, i=1,2,…, to encrypt a long message sequence.
B_i^{-1}	Use B in a decryption application (D-application) at a i'th block slot, i=1,2,…, to decrypt a long message sequence.
P^T	A message sequence of all the plaintext block data.
C^T	A message sequence of all the ciphertext block data.
P_i	An i'th block data in a session of task. $P_i=D\text{-}application[C_i]=B^{-1}[C_i]$
C_i	An i'th block data in a session of task. $C_i=E\text{-}application[P_i]=B[P_i]$
S_i	An i'th output State (or keyStream) of the mode.

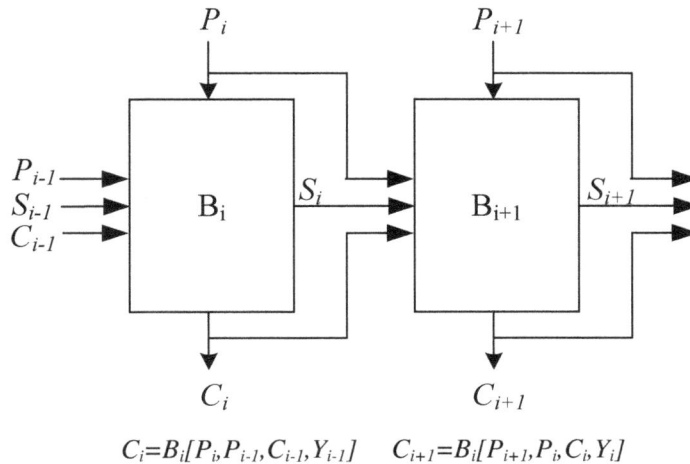

$$C_i=B_i[P_i,P_{i-1},C_{i-1},Y_{i-1}] \qquad C_{i+1}=B_i[P_{i+1},P_i,C_i,Y_i]$$

(a) The encryption operations

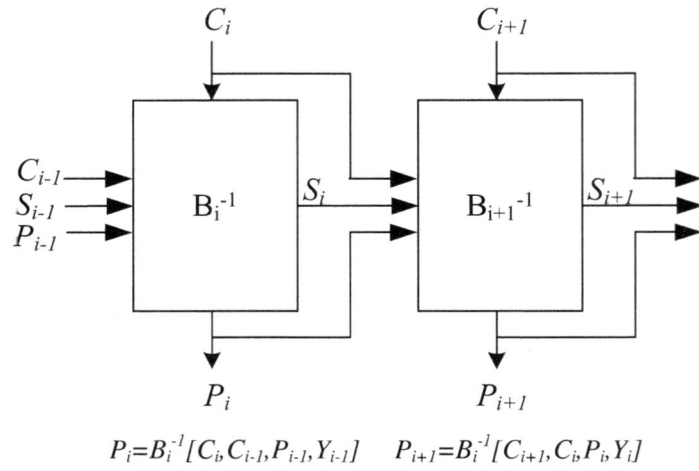

$$P_i=B_i^{-1}[C_i,C_{i-1},P_{i-1},Y_{i-1}] \qquad P_{i+1}=B_i^{-1}[C_{i+1},C_i,P_i,Y_i]$$

(b) The decryption operations

Figure 8. Real-time Changing Operational Mode

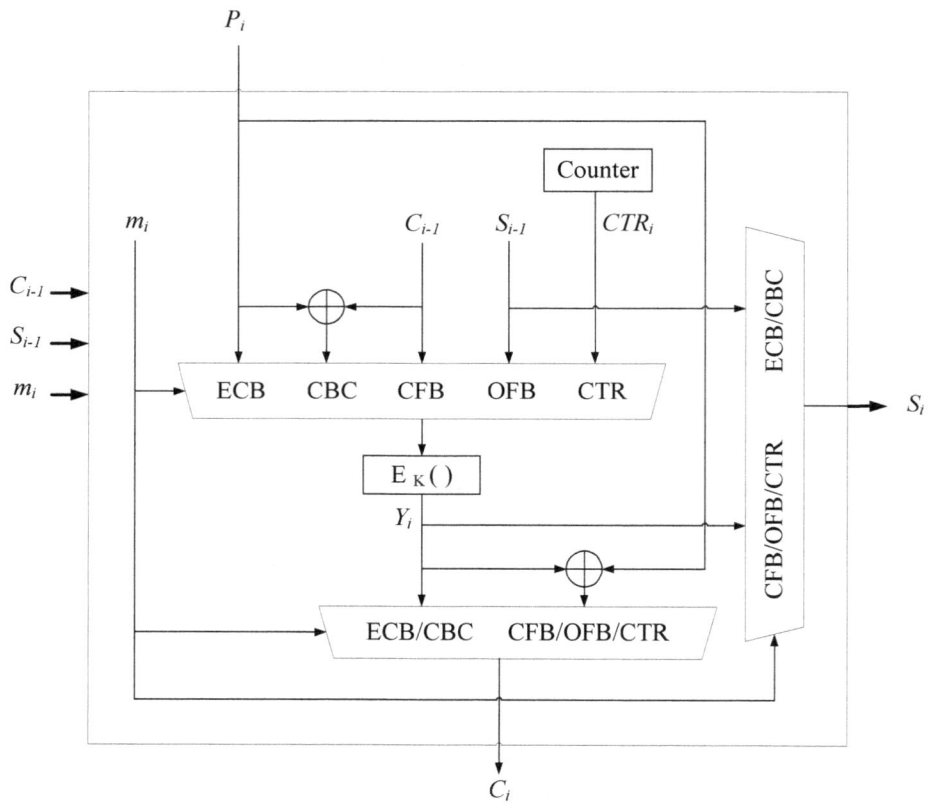

(a) Crypto module $\mathbf{B_i}$

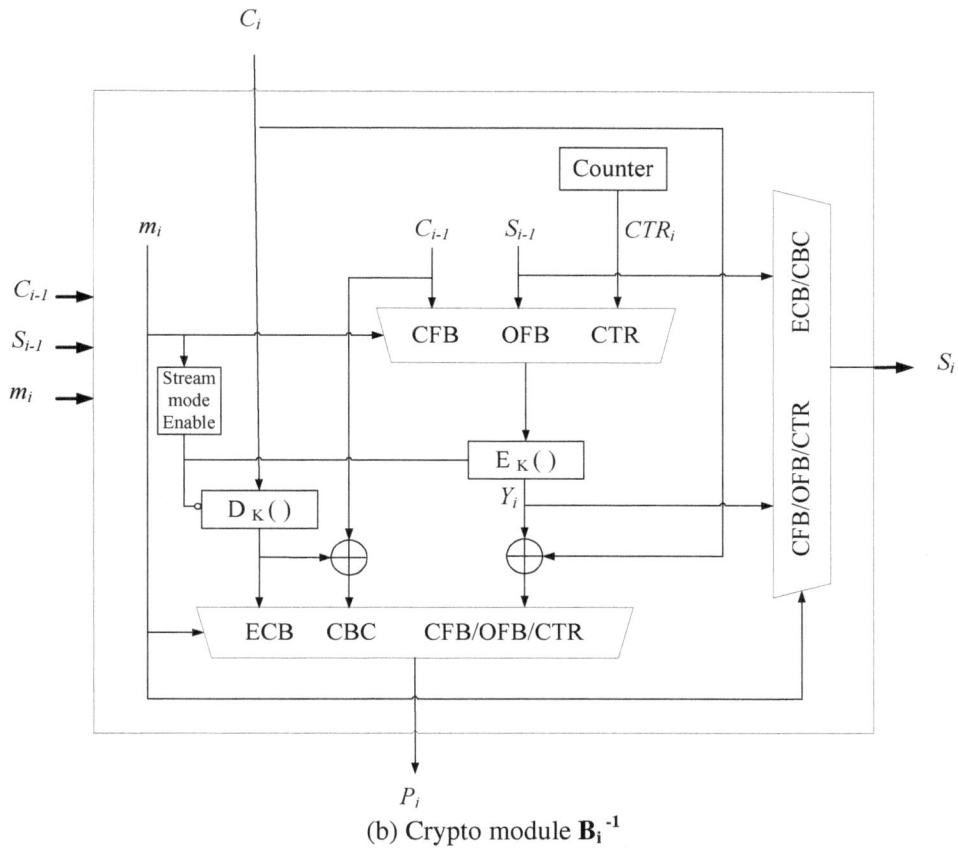

(b) Crypto module $\mathbf{B_i}^{-1}$

Figure 9. System Architecture of Proposed Crypto Module

6. CONCLUSIONS

The design of a real-time hopping mode of operation is proposed in this paper. In this manuscript, the author proposes a seamless changing technique for block ciphers. The changing technique facilitates the smooth implementation of proposed system architecture of dynamic changeable and low-resource hardware, which is appropriate for mobile devices such as a smartphone. According to most of the content in mobile communication are products from social applications or multimedia applications. In the future, people share their own video programs frequently. The author provide a common solution for multi-mode, which is highly suitable for recent block cipher computing using several resources in future various environments, especially a large amount data such as video streaming.

REFERENCES

[1]   iDevice. http://programming4.us/multimedia/4782.aspx

[2]   SSL 3.0 Specification. http://wp.netscape.com/eng/ssl3/

[3]   IPSec Working Group. http://www.ietf.org/html.charters/ipseccharter.html

[4]   Block cipher modes of operation.
        http://en.wikipedia.org/wiki/Block_cipher_modes_of_operation

[5]   S. Guilley, P. Hoogvorst, and R. Pacalet, (2007) "A Fast Pipelined Multi-Mode DES Architecture Operating in IP Representation," *The VLSI Journal*, Vol. 40, No. 4, pp.479-489.

[6]   L. Luo, Z.G. Qin, and J. Wang, (2009) "The Intelligent Conversion for Different Layers' Block Ciphers," *ICIC Express Letters*, Vol. 3, No. 1, pp.73–77.

[7]   C.P. Young, C.C. Chia, Y.B. Lin, and L.B. Chen, (2011) "Fast multi-cipher transformation and its implementation for modern secure protocols," *IJICIC*, Vol. 7, No. 8, pp.4941–4954.

[8]   K.T. Huang, J.H. Chiu, and S.S. Shen, (2013) "A Novel Structure with Dynamic Operation Mode for Symmetric-Key Block Ciphers," *International Journal of Network Security & Its Applications*, Vol. 5, No. 1, pp.17-36.

[9]   National Institute of Standards and Technology (NIST), NIST. gov - Computer Security Division - Computer Security Resource Center, "Recommendation of block cipher security methods and Techniques," NIST SP800-38.

[10]   International Organization for Standardization (ISO), "Information Technology-Security Techniques-Modes of Operation for. an n-bit Block Cipher," ISO/IEC 10116.

[11]   William Stallings (2003). *Cryptography and Network Security: Principles and Practices 3rd Edition*. Pearson Education. ISBN 0-13-111502-2.

[12]   Alfred J. Menezes, Paul C. van Oorschot and Scott A. Vanstone (1996). *Handbook of Applied Cryptography*. CRC Press. ISBN 0-8493-8523-7.

[13]   G.J. Simmons (1999). *Contemporary cryptology: The science of information integrity*. Wiley. ISBN 0-7803-5352-8.

[14]   J. Daemen and V. Rijmen (2002). *The Design of Rijndael: AES - the advanced encryption standard*. Springer Verlag.

[15]   S. Mangard, M. Aigner and S. Dominikus, (2003) "A highly regular and scalable AES hardware architecture," *IEEE Transaction on Computers*, Vol. 52, No. 4, pp. 483–491.

[16]   D. Carlson, D. Brasili, A. Hughes, A. Jain, T. Kisezly, P. Kodandapani, A. Vardharajan, T. Xanthopoulos, and V. Yalala, (2003) "A high performance SSL IPsec protocol security processor," *Proc. ISSCC'03*.

[17]   A. Hodjat, P. Schaumont, and I. Verbauwhede, (2004) "Architectural design features of a programmable high throughput AES coprocessor," *Proc. ITCC'04*, pp.498–502.

[18]   A. Satoh, S. Morioka, K. Takano, and S. Munetoh, "A compact Rijndael hardware architecture with S-box optimization," *Proc. ASIACRYPT 2001*, Lect Notes Comput Sci 2248, pp.239–254.

[19]   P. Saravanan, N. RenukaDevi, G. Swathi, and P. Kalpana, (2011) "A high-throughput ASIC implementation of configurable AES processor," *IJCA*, NSC(3), pp.1–6.

[20]   Open Mobile Alliance, Approved Version 2.1, 2008.

[21]   K.T. Huang, J. Stu, S.C. Chou, and J.H. Chiu, (2011) "OTP-based hybrid cipher using for mobile e-Learning," *Proc. 2nd World Congress on Computer Science and Information Engineering*, pp. 127–134.

[22]   M. Feldhofer and J. Wolkerstorfer, (2007) "Strong crypto for RFID tags - A comparison of low-power hardware implementations," *Proc. IEEE ISCAS'07*, pp. 1839–1842.

[23]   M. Grangetto, A. Grosso, and E. Magli, (2004) "Selective encryption of JPEG 2000 images by means of randomized arithmetic coding," *Proc. IEEE International Workshop on Multimedia Signal Processing*, pp. 347–350.

[24]   A.M. Alattar and G.I. Al-Regib, (1999) "Evaluation of selective encryption techniques for secure transmission of MPEG video bit-streams," *Proc. IEEE ISCAS'99*, pp340–343.

[25]   L. Qiao and K. Nahrstedt, (1998) "Comparsion of MPEG encryption algorithms," *Computers and Graphics*, Vol. 22, No. 4, pp. 1–21.

[26]   I. Agi and L. Gong, (1996) " An empirical study of MPEG video transmissions," *Proc. Internet Society Symposium on Network and Distributed System Security*, pp. 137–144.

[27]   L. Wu, C. Weaver, and T. Austin, (2001) "CryptoManiac: A Fast Flexible Architecture for Secure Communication," *Proc. IEEE Int. Symp. Comput. Archit.*, pp. 110–119.

[28]   S. Laovs, A. Priftis, P. Kitsos, and O. Koufopavlou, (2003) "Reconfigurable crypto process design of encryption algorithms operation modes methods and FPGA integration," *Proc. IEEE Int. Conf. MWSCAS*, pp. 811–814.

4

PERFORMANCE COMPARISON OF TOPOLOGY AND POSITION BASED ROUTING PROTOCOLS IN VEHICULAR NETWORK ENVIRONMENTS

Akhtar Husain[1], Ram Shringar Raw[2], Brajesh Kumar[1], and Amit Doegar[3]

[1] Department of Computer Science & Information Technology, MJP Rohilkhand University, Bareilly, India
husainakhtar@yahoo.com, bkumar@mjpru.ac.in
[2] School of Computer and Systems Sciences, Jawaharlal Nehru University, New Delhi, India,
rsrao08@yahoo.in
[3] Department of Computer Science, NITTTR, Chandigarh, India
amit@nitttrchd.ac.in

ABSTRACT

Due to mobility constraints and high dynamics, routing is a more challenging task in VANETs. In this work, we evaluate the performance of the routing protocols in vehicular network environment. The objective of this work is to assess the applicability of these protocols in different vehicular traffic scenarios. Both the position-based and topology-based routing protocols have been considered for the study. The topology-based protocols AODV and DSR and position-based protocol LAR are evaluated in city and highway scenarios. Mobility model has significant effects on the simulation results. We use Intelligent Driver Model (IDM) based tool VanetMobiSim to generate realistic mobility traces. Performance metrics such as packet delivery ratio, throughput, and end-to-end delay are evaluated using NS-2. Simulation results shows position based routing protocols gives better performance than topology based routing protocols.

KEYWORDS

VANET, Routing Protocols, AODV, DSR, LAR, City Scenario, Highway Scenario, Intelligent Driver Model.

1. INTRODUCTION

Vehicular ad hoc networks are important technology for future developments of vehicular communication systems. Such networks composed of moving vehicles, are capable of providing communication among nearby vehicles and the roadside infrastructure. Modern vehicles are equipped with computing devices, event data recorders, antennas, and GPS receivers making VANETs realizable. VANETs can be used to support various functionalities such as vehicular safety [1], reduction of traffic congestion, office-on-wheels, and on-road advertisement. Most nodes in a VANET are mobile, but because vehicles are generally constrained to roadways, they have a distinct controlled mobility pattern [2]. Vehicles exchange information with their neighbors and routing protocols are used to propagate information to other vehicles. Some important characteristics that distinguish VANETs from other types of ad hoc networks include:

* High mobility that leads to extremely dynamic topology.
* Regular movement, restricted by both road topologies and traffic rules.

- Vehicles have sufficient power, computing and storage capacity.
- Vehicles are usually aware of their position and spatial environment.

Unlike conventional ad hoc wireless networks a VANET not only experiences rapid changes in wireless link connections, but may also have to deal with different types of network topologies. For example, VANETs on freeways are more likely to form highly dense networks during rush hours, while VANETs are expected to experience frequent network fragmentation in sparsely populated rural freeways or during late night. High and restricted node mobility, sort radio range, varying node density makes routing a challenging job in VANETs.

A number of routing protocols have been proposed and evaluated for ad hoc networks. Some of these are also evaluated for VANET environment, but most of them are topology based protocols. Position based protocols (like Location Aided Routing (LAR) [3] [4]), which require information about the physical position of the participating nodes, have not been studied that much and require more attention. The results of performance studies heavily depend on chosen mobility model. The literature shows that the results of many of performance studies are based on mobility models where nodes change their speed and direction randomly. Such models cannot describe vehicular mobility in a realistic way, since they ignore the peculiar aspects of vehicular traffic such as vehicles acceleration and deceleration in the presence of nearby vehicles, queuing at roads intersections, impact of traffic lights, and traffic jams. These models are inaccurate for VANETs and can lead to erroneous results.

In this paper performance evaluation of Ad Hoc On Demand Distance Vector (AODV) [5], Dynamic Source Routing (DSR) [6], and LAR in city and highway traffic environments under different circumstances is presented. To model realistic vehicular motion patterns, we use the Advanced Intelligent Driver Model which is the extension of Intelligent Driver Model (IDM) [7]. We evaluate the performance of AODV, DSR, and LAR in terms of Packet Delivery Ratio (PDR), throughput, and end-to-end delay.

The rest of the paper is organized as follows: Section 2 presents related work, while section 3 briefly depicts the three candidates' protocols. Mobility model for VANET is presented in section 4. Section 5 presents simulation methodology and describes reported results. Finally, in section 6, we draw some conclusion remarks and outline of future works.

2. RELATED WORK

Routing protocols can be classified in two broad categories: topology based and position based routing protocols [8] [9]. Topology based approaches, which are further divided into two subcategories: reactive and proactive, use information about links to forward the packets between nodes of the network. Prominent protocols of this category are AODV, and DSR. Position-based routing (e.g. LAR) requires some information about the physical or geographic positions of the participating nodes. In this protocol the routing decision is not based on a routing table but at each node the routing decision is based on the positions of its neighboring nodes and the position of the destination node.

Several studies have been published comparing the performance of routing protocols using different mobility models, and different traffic scenarios with different performance metrics. A paper by Lochert et al. [10] compared AODV, DSR, and Geographic Source Routing (GSR) in city environment. They show that GSR which combines position-based routing with topological knowledge outperforms both AODV and DSR with respect to delivery rates and latency. A study by Jaap et al. [11] examined the performance of AODV and DSR in city traffic scenarios. Another study presented by Juan Angel Ferreiro-Lage et al. [12] compared AODV and DSR protocols for vehicular networks and concluded that AODV is best among the three protocols. LAR is described in [3] is to reduce the routing overhead by the use of position information. Position information will be used by LAR for restricting the flooding to a certain area called request zone. Authors found that LAR is more suitable for VANET.

3. DESCRIPTION OF ROUTING PROTOCOLS

The topology based routing protocols AODV and DSR are the most studied routing protocols as evident from section 2. The position-based routing protocols need more attention. We have chosen LAR, which is a position-based routing protocol for the performance comparison with AODV and DSR. In this section a brief description of these routing protocols is given.

3.1 Dynamic Source Routing Protocol (DSR)

In DSR, when a source node wants to send a packet to a particular destination it checks route cache first. If it does not find any route, it initiates the route discovery protocol. Route discovery involves broadcasting a route request packet. The route request packet contains the address of the source and the destination, and a unique identification number. Each intermediate node checks whether it knows of a route to the destination. If it does not, it appends its address to the route record of the packet and forwards the packet to its neighbors. If the route discovery is successful then the source host, which requested route discovery, receives a route reply packet. Route reply packet lists all the network hops through which target node may be reached. To limit the number of route requests propagated, a node processes the route request packet only if it has not already seen the packet and its own address is not present in the route record of the packet. A route reply is generated when either the destination or an intermediate node with current information about the destination receives the route request packet.

3.2 Ad Hoc On Demand Distance Vector (AODV)

In AODV, whenever a source node has to communicate with a destination node such that it has no routing information in its table, it first initiates route discovery process. The node broadcasts a route request (RREQ) packet to all its neighbors. The route request packets contains source address, source sequence number, broadcast ID, destination address, destination sequence number and hop count. The source address and broadcast ID uniquely identifies a RREQ. The source sequence number is for maintaining the freshness of information about the reverse route to the source. The destination sequence number specifies how fresh a route to the destination must be before it can be accepted by the source. If a neighbor knows the route to the destination, it replies with a route reply control message RREP that propagates through the reserve path. Otherwise, the neighbor will re-broadcast the RREQ until an active route is found or the maximum number of hops is reached.

3.3 Location Aided Routing (LAR)

The goal of Location-Aided Routing (LAR) described in [3] is to reduce the routing overhead by the use of location information. Position information will be used by LAR for restricting the flooding to a certain area called request zone. As a consequence, the number of route request messages is reduced. Instead of flooding the whole network with route discovery message, this protocol send messages to a subset of nodes from whom the probability of finding route is very high. In LAR, an expected zone is defined as a region that is expected to hold the current location of the destination node. During route discovery procedure, the route request flooding is limited to a request zone, which contains the expected zone and location of the sender node.

4. MOBILITY MODEL

Simulation is a popular approach for evaluating routing protocols; however the accuracy of the results depends on the mobility used. Mobility model has significant effects on the simulation results. Random Waypoint (RWP) model, which is widely used for MANET simulations, is unsuitable for VANET simulations as the mobility patterns underlying an inter-vehicle network are quite different. In order to model realistic vehicular motion patterns The mobility model used for studying VANETs must reflect as close as possible, the behavior of vehicular traffic. Both macro-mobility and micro-mobility aspects need to be considered jointly [13].

4.1 Macro-Mobility

Macro-mobility includes all the macroscopic aspects which influence vehicular traffic. The concept of macro-mobility not only takes into account the road topology, but also includes the road structure and characteristics such as unidirectional or bidirectional traffic flows, single or multi-lane, intersection crossing rules, speed limits, vehicle-class based restriction, and the presence of stop signs, traffic lights, etc. The length of streets, the frequency of intersections, and the density of buildings can greatly affect the important mobility metrics such as the speed of the vehicles, or their density over the simulated map.

Path selection strategy is mainly reflected in macro scope. Existing mobility models do not place enough attention on path selection strategies. They simply use shortest path algorithm to compute every node's shortest path. But this approach is not suitable to inter-vehicle communication environment. Vehicles equipped with GPS can get information about local traffic, congestion, and road situation and may choose different path according to their roles even if they have same destination.

4.2 Micro-Mobility

The vehicular micro-mobility model includes all aspects related to an individual vehicle's speed and acceleration modeling. When moving along the road, there are typically four state of a vehicle's motion: stop, accelerate, decelerate and move at a constant speed. All vehicles move with changing its motion state, but the changing strategies are different [14]. Acceleration strategy is mainly responsible for speed differences, while stop strategy mainly related to travelling patterns. A vehicle's stop may attribute to reactive reasons or proactive reasons. When a vehicle meets a congestion or accident which blocks it from moving ahead, it is forced to stop. However, a public transport vehicle must stop at every stop for several seconds. These two stop behaviors consume different time.

Micro-mobility refers to drivers' individual behavior, when interacting with other drivers or with the road infrastructure: travelling speed in different conditions; acceleration, deceleration, overtaking criteria, behavior in the presence of road intersections and traffic signs, and general driving attitude. The micro-mobility description plays the main role in the generation of realistic vehicular movements, as it is responsible for effects such as smooth speed variation, vehicles queues, traffic jams and over takings.

4.3 The Intelligent Driver Model

The Intelligent Driver Model (IDM) [15] is a car-following model that characterizes drivers' behavior depending on their front vehicles.Vehicles acceleration/deceleration and its expected speed are determined by the distance to the front vehicle and its current speed. Moreover, it is also possible to model the approach of vehicles to crossings. Another advantage of the IDM is that it uses a small set of parameters that which can be evaluated with the help of real traffic measurements. The instantaneous acceleration of a vehicle is computed according to the following equation:

$$\frac{dv}{dt} = a\left[1 - \left(\frac{v}{v_0}\right) - \left(\frac{S^*}{S}\right)^2\right]$$

(1)

Where v is the current speed of the vehicle, v_0 is the desired velocity, S is the distance from the preceding vehicle and S^* is the desired dynamical distance to the vehicle/obstacle, which computed with the help of equation (2).

$$S^* = S_0 + \left[vT + \left(\frac{v.\Delta v}{2\sqrt{ab}}\right)\right]$$

(2)

Desired dynamical distance S^* is a function of jam distance S_0 between two successive vehicles, the minimum safe time headway T, the speed difference with respect to front vehicle velocity Δv, maximum acceleration a, and maximum deceleration b.

VanetMobiSim developed by J. Harri et al [16] [17], extends IDM and adds two new microscopic mobility models Intelligent Driver Model with Intersection Management (IDM-IM) and Intelligent Driver Model with Lane Changing (IDM-LC). IDM-IM adds intersection handling capabilities to the behavior of vehicles driven by the IDM. Moreover, IDM-IM models two different intersection scenarios: a crossroad regulated by the stop signs, and a road junction ruled by traffic lights. IDM-LC extends the IDM-IM model with the possibility for the vehicles to change lane and overtake each others.

5. EXPERIMENTS AND EVALUATIONS

Extensive simulations have been carried out to evaluate and compare the performances of LAR, AODV, and DSR in VANETs by using the network simulator NS-2 in its version 2.32. It is freely available and widely used for research in mobile ad hoc networks. The movements of nodes are generated using VanetMobiSim tool [18]. The awk programming is used to analyze the simulation results. It is assumed that every vehicle is equipped with GPS and can obtain its current location.

5.1 System Model

Vehicles are deployed in a 1000m *1000m area. A Manhattan grid like road network is assumed to have eight vertically and horizontally oriented roads and 16 crossings. The vehicle moves and accelerates to reach a desired velocity. When a vehicle moves near other vehicles, it tries to overtake them if road includes multiple lanes. If it cannot overtake it decelerate to avoid the impact.

Table 1. Mobility model parameters

Parameter	Value
Maximum Acceleration	0.9 m/s^2
Maximum Deceleration	0.5 m/s^2
Maximum safe deceleration	4 m/s^2
Vehicle Length	5 m
Traffic light transition	10s

Lane change threshold	$0.2\ m/s^2$
Politeness	0.5
Safe headway time	1.5 s
Maximum congestion distance	2 m

When a vehicle is approaching an intersection, it first acquires the state of the traffic sign. If it is a stop sign or if the light is red, it decelerates and stops. If it is a green traffic light, it slightly reduces its speed and proceeds to the intersection. The other mobility parameters are given in table 1.

Table 2. Simulation parameters

Parameter	Value
MAC and Channel Type	IEEE 802.11,Wireless Channel
Antenna type	Omni directional
Simulation time	1500 seconds
Simulation area	1000m x 1000m
Transmission range	250m
Node speed	30 km/hr and 100 km/hr
Traffic type	CBR
Data payload	512 bytes/packet
Packet rate	4 packets/sec
Node pause time	20 s
Bandwidth	2 Mbps
Mobility model	IDM based
Interface queue type	Drop Tail/Priority Queue, and CMU Priori Queue
Interface queue length	50 packets
Receiving threshold	Two-Ray Rx Power (Radio Range)
Sensing threshold	Two-Ray Rx Power (Radio Range + 5 m)
No. of vehicles	10 to 80

Vehicles are able to communicate with each other using the IEEE 802.11 DCF MAC layer. The transmission range is taken to be 250 meters. The traffic light period is kept constant at 60 seconds. Simulations are repeated varying the speed, that is, 30 km/h (city) and 100 km/h (highway) and varying the node density. The other simulation parameters are given in Table 2. Only one lane case is taken for city scenario. However, in highway scenario, first one lane case is considered and later it is generalized to multiple lanes. The other simulation parameters are given in Table 2.

5.2 Results and Discussion

The protocols are evaluated for packet delivery ratio, throughput, and average end-to-end delay at varying node densities (10 to 80 vehicles). Hereafter the terms node and vehicle are used interchangeably.

5.2.1 Packet Delivery Ratio (PDR)

Packet delivery ratio is defined as the ratio of data packets received by the destinations to those generated by the sources. Mathematically, it can be defined as:

$$PDR = \frac{S_a}{S_b}$$

(3)

Where, S_a is the sum of data packets received by the each destination and S_b is the sum of data packets generated by the each source.

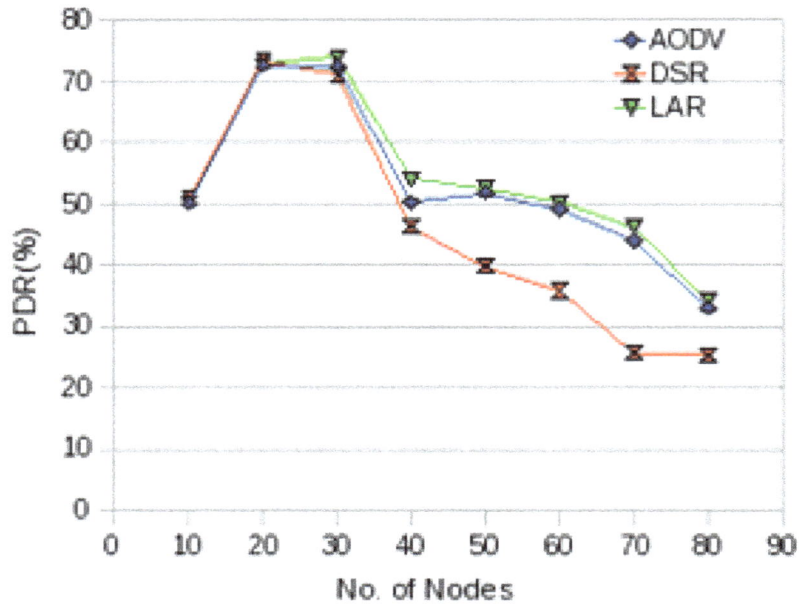

(a) City Scenario using IDM-IM

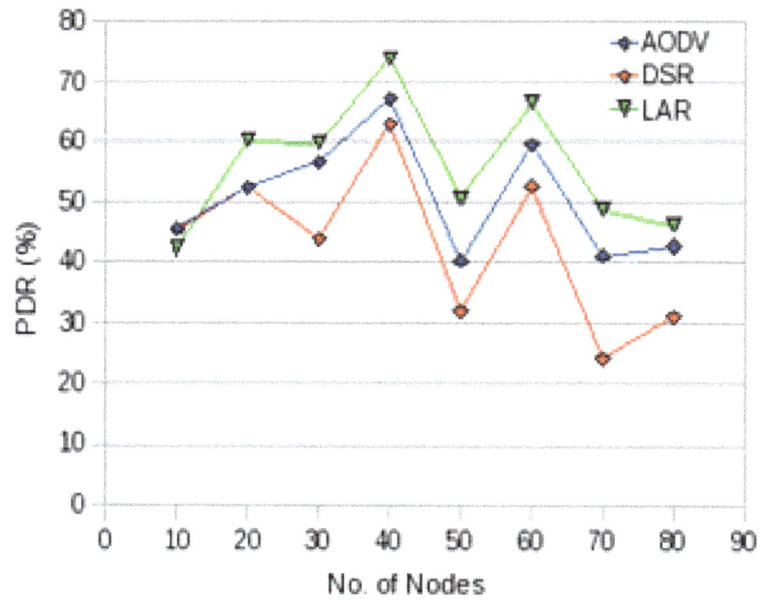

(b) Highway Scenario using IDM-IM

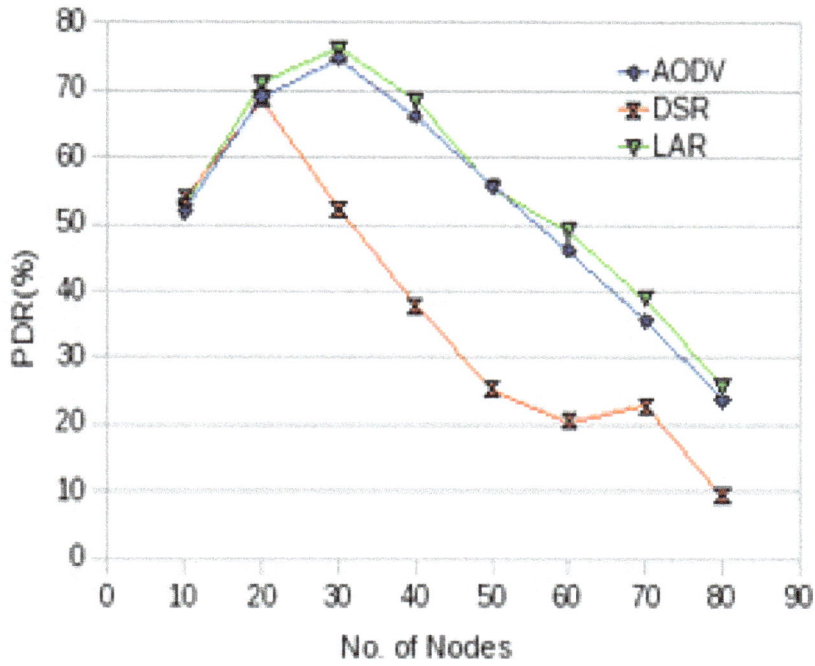

(c) Highway Scenario using IDM-LC (lane = 2)

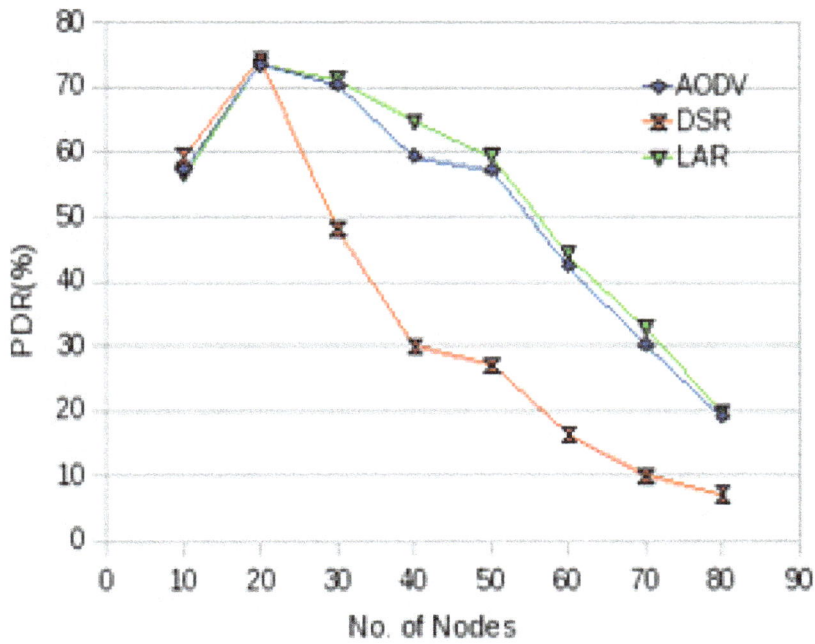

(d) Highway Scenario using IDM-LC (lane = 4)

Figure 1. PDR as a function of node density

Figure 1 depicts the fraction of data packets that are successfully delivered during simulations time versus the number of nodes. In city scenario all the four protocols exhibit good performance for low node densities as shown in Figure 1(a) and none of the protocols clearly outperforms the others. But with the increase in density, performance of the protocols decreases. It can be observed that the performance of the DSR reduces drastically while LAR is slightly better among the three. In highway scenario all the protocols show relatively better performance as depicted by Figure 1(b) and location based protocol clearly outperform the topology-based protocols with LAR exhibiting best results. Though, PDR again decreases with increasing density, but it is not that much low as observed in city scenario. Figure 1(c) and Figure 1(d) show that performance of the protocols degrades in highway scenario with lane changing (LC). It further slightly degrades with increase in the number of lanes. Again DSR is the worst and LAR has a slight edge over the other protocols.

5.2.2 Throughput

Number of bits delivered successfully per second to the destination. It is the sum of bits received successfully by all destinations. It is represented in kilo bits per second (kbps). Mathematically, it can be defined as:

$$Throughput = \frac{N}{1000}$$

(4)

Where N is the number of bits received successfully by all destinations.

Figure 2 illustrates the variation of throughput by varying node densities. Results show that throughput of all the candidate protocols clearly increases between densities of 10 and 30 nodes. This behavior results from the network connectivity. When density is low, in many cases it may not be possible to establish a connection between source and destination pair.

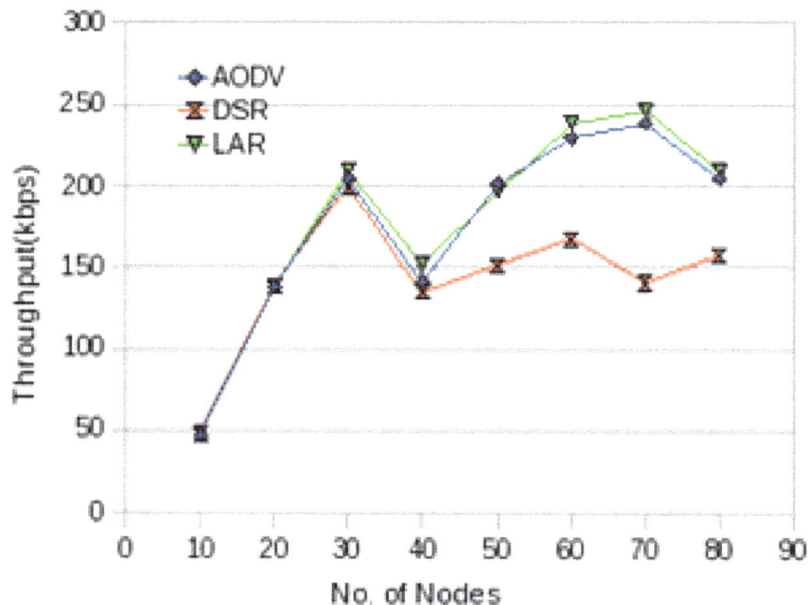

(a) City scenario using IDM-IM

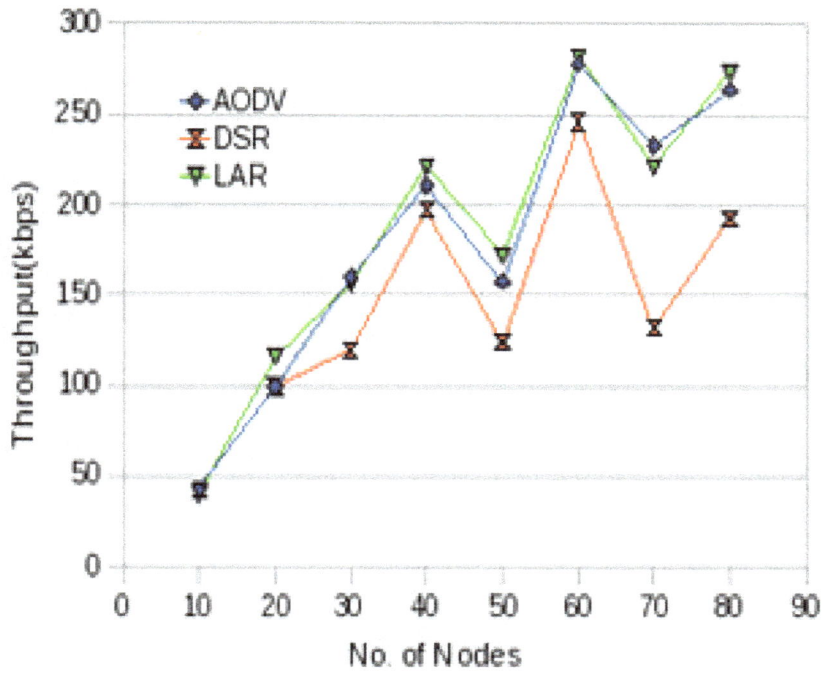

(b) Highway scenario using IDM-IM

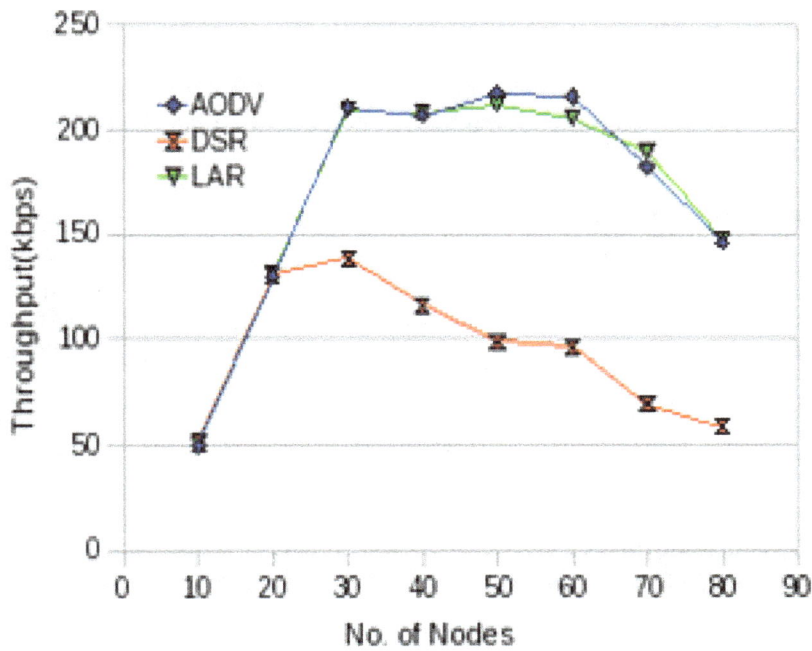

(c) Highway scenario using IDM-LC (lane = 2)

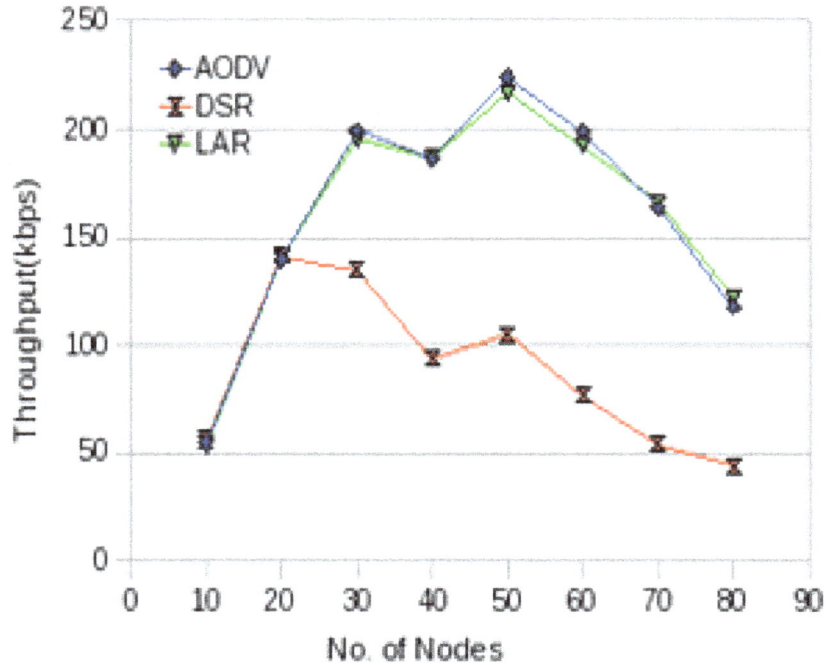

(d) Highway scenario using IDM-LC (lane = 4)

Figure 2. Throughput vs node densities

Throughput improves when connectivity is better. At higher densities, the throughput of all the protocols is reduced. It is due to increased routing overhead when wireless channel is shared by more and more vehicles. Again DSR is not coming good and LAR clearly outperforms the other protocols, especially beyond density of 30 nodes.

5.2.3 Average End-to-End Delay (EED)

The average time it takes a data packet to reach the destination. This includes all possible delays caused by buffering during route discovery latency, queuing at the interface queue, retransmission delay at MAC, and propagation delay. This metric is calculated by subtracting time at which first packet was transmitted by source from time at which first data packet arrived to destination. Mathematically, it can be defined as:

$$\text{Average EED} = \frac{S}{N}$$

(5)

Where S is the sum of the time spent to deliver packets for each destination, and N is the number of packets received by the all destination nodes.

Finally, Figure 3 summarizes the variation of the average latency by varying node density. Average latency increases with increasing the number of nodes. DSR consistently presents the highest delay. This may be explained by the fact that its route discovery process takes a quite long time compared to other protocols. LAR has the lowest delay though compared to DSR and AODV it is low only slightly. In city scenario the location-based protocol exhibit lower delay compared to AODV especially at higher densities. Delay of all the protocols clearly increases in highway scenario with lane changing.

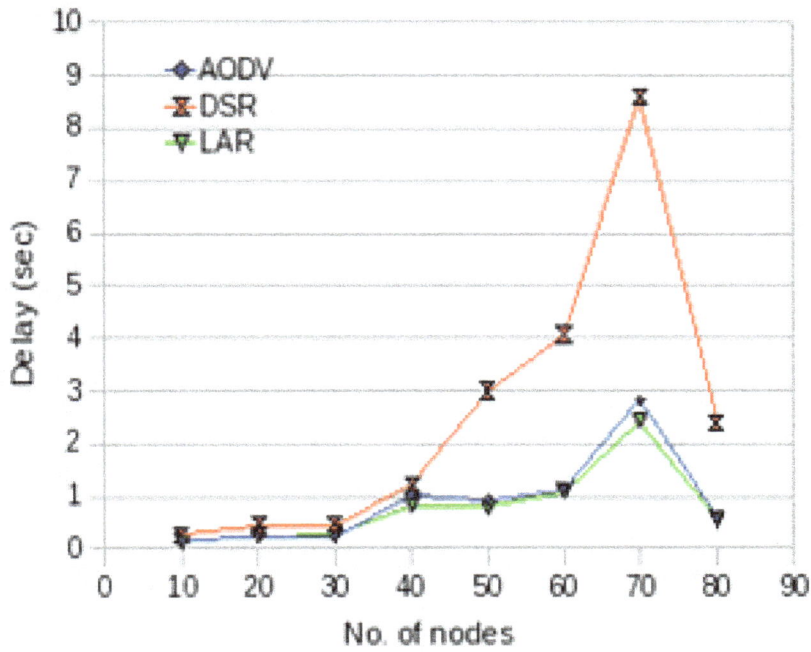

(a) City scenario using IDM-IM

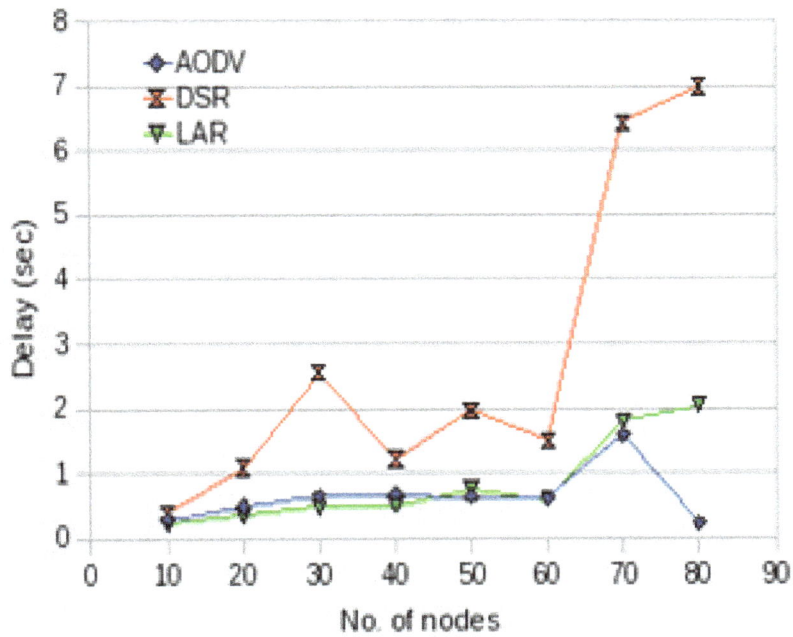

(b) Highway scenario using IDM-IM

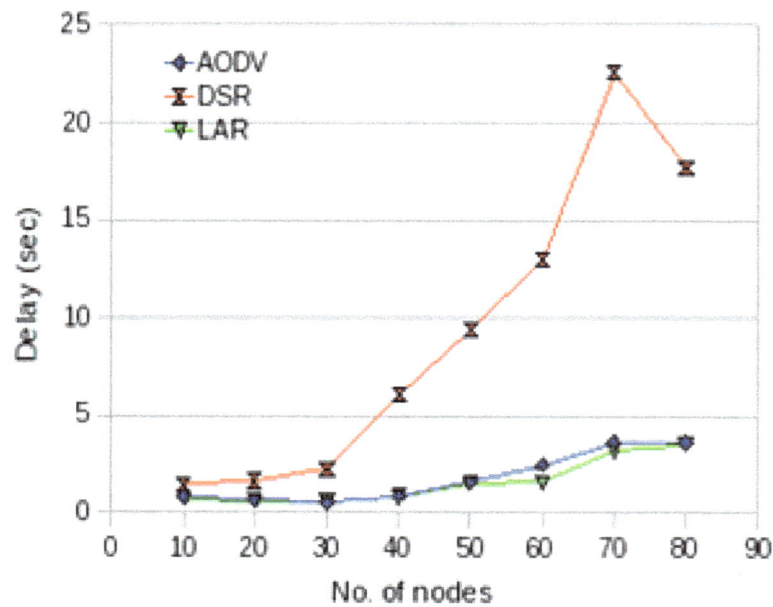

c) Highway scenario using IDM-LC (lane = 2)

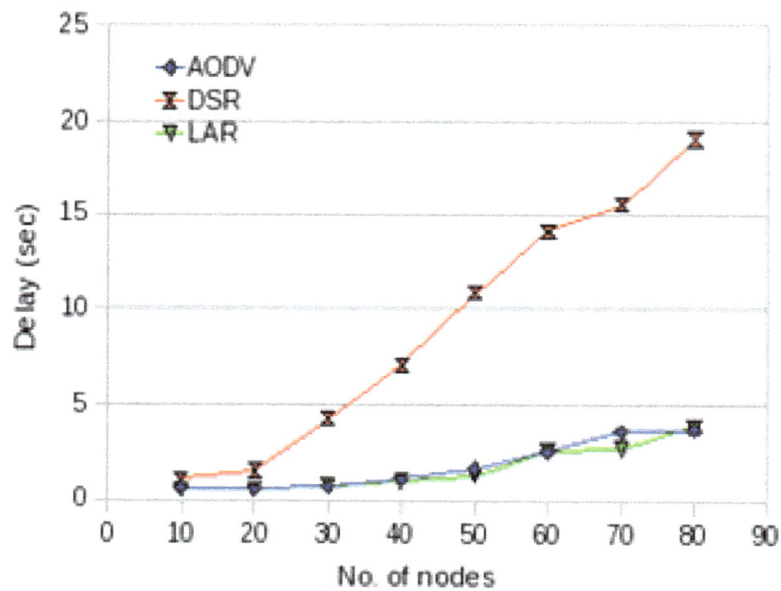

(d) Highway scenario using IDM-LC (lane = 4)

Figure 3. Delay as a function of node density

Overall, it can be concluded that LAR showed the best performance in the simulated scenarios with highest PDR. Moreover, it has the highest throughput and shown the lower delays also. It can be said that DSR is not a suitable protocol for VANET whether it is city scenario or highway scenario. AODV also showed the good performance with better PDR and throughput and lower delays.

6. CONCLUSION AND FUTURE WORK

In this paper, performance of three routing protocols AODV, DSR, and LAR were evaluated for vehicular ad hoc networks in city and highway scenarios. We used Advanced Intelligent Driver Model to generate realistic mobility patterns. The three protocols were tested against node density for various metrics. It is found that position based routing protocol (LAR) outperforms topology based routing protocols (DSR and AODV) in different VANET environment. For most of the metrics LAR has the better performance.

Overall, it can be concluded that position based routing protocol gives better performance than topology based routing protocols in terms of packet delivery ratio, throughput, and end-to-end delay for both the vehicular traffic scenarios. For future work, we want to study the impact of traffic signs, traffic lights and transmission range on the performance of the routing protocols. Future work will also include the evaluation of other position based routing protocols as they are more suitable in vehicular traffic environment.

REFERENCES

[1]   X. Yang, J. Liu, F. Zhao, and N. Vaidya, (2004) "A Vehicle-to Vehicle Communication Protocol for Cooperative Collision Warning", Int'l Conf. On Mobile and Ubiquitous Systems: Networking and Services (MobiQuitous 2004), pp. 44-45.

[2]   J. Bernsen, and D. Manivannan, (2009) "Unicast routing protocols for vehicular ad hoc networks", A critical comparison and classification, Elsevier Journal of Pervasive and Mobile Computing,vol. 5, issue 1, pp. 1-18.

[3]   Y. Ko, and N. Vaidya, (2002) "Location Aided Routing in Mobile Ad Hoc Networks", ACM journal of Wireless Networks, vol. 6, no. 4, pp 307-321.

[4]   T. Camp, J. Boleng, B. Williams, L. Wilcox, and W. Navidi, (2002) "Performance Comparison of Two Locations Based Routing Protocols for Ad Hoc Networks", Proceedings of the IEEE INFOCOM,vol.3, pp. 1678-1687.

[5]   C. E. Perkins, and E. M. Royer, (1999) "Ad hoc on-demand distance vector routing", Proceedings of the 2nd IEEE Workshop on Mobile Computing Systems and Applications, New Orleans, I.A. pp. 90-100.

[6]   D. B. Johnson, D. A. Maltz, and J. Broch, (1985) "Dynamic source routing protocol for wireless ad hoc network", IEEE Journal on Selected Areas in Communication, Vol. 3, Issue 3, pp.431-439.

[7]   J. Haerri, F. Filali, and C. Bonnet, (2006) "Performance comparison of AODV and OLSR in VANETs urban environments under realistic mobility patterns", 5th annual IFIP Mediterranean Ad Hoc Networking Workshop (Med-Hoc-Net 2006), Lipari, Italy.

[8]   M. Mauve, J. Widmer, and H. Hartenstein, (2001) "A survey on position-based routing in mobile ad hoc networks", IEEE Network Magazine, Vol. 15, Issue No.6, pp. 30-39.

[9]   X. Hong, K. Xu, and M. Gerla, (2002) "Scalable Routing Protocols for Mobile Ad Hoc Networks", IEEE Network Magzine, vol. 16, no. 4, pp 11-21.

[10]   C. Lochert, H. Hartenstein, and J. Tian, (2003) "A Routing strategy for vehicular ad hoc networks in city environments", Proceedings of IEEE Intelligent Vehicles Symposium, Columbus, USA, pp. 156-161.

[11]   S. Jaap, M. Bechler, and L. Wolf, (2005) "Evaluation of routing protocols for vehicular ad hoc networks in city traffic scenarios", Proceedings of the 5th International Conference on Intelligent Transportation Systems Telecommunications (ITST),Brest, France.

[12]   J. A. Ferreiro-Lage, C. P. Cristina Pereiro Gestoso, O. Rubinos, and F. A. Agelet, (2009) "Analysis of Unicast Routing Protocols for VANETs", 5th Intl. Conference on Networking and Services, Valencia Spain, pp 518-521.

[13]   S. Xi, and X. Li, (2008) "Study of the Feasibility of VANET and its Routing Protocols", Proceedings of 4th Int. Conference, WiCOM'08, Dalian, pp 1-4.

[14]   M. Bakhouya, J. Gaber, and M. Wack, (2009) "Performance evaluation of DREAM protocol for Inter-Vehicle Communication", 1st Intl. Conference on Wireless Communications, Vehicular Tech, Information Theory, and Aerospace & Electronics Systems Technology, Wireless VITAE 2009, Aalborg, pp 289-293.

[15]   M. Trieber, A. Hennecke, and D. Helbing, (2000) "Congested traffic states in empirical observations and microscopic simulations", Phys. Rev. E 62, Issue 2.

[16]   J. Harri, and M. Fiore, (2006) "VanetMobiSim- vehicular ad hoc network mobility extension to the CanuMobiSim framework", Manual, Institute Eurecom/Politecnico di Torino, Italy.

[17]   The Network Simulator NS-2 (2006), URL: http://www.isi.edu/nsnam/ns/.

[18]   VanetMobiSim, URL: http://vanet.eurecom.fr/.

5

Chaining Algorithm and Protocol for Peer-to-Peer Streaming Video on Demand System

Hareesh.K[1]   and   Manjaiah D.H[2]

[1]Research Scholar, Jawaharlal Nehru Technological University, Anantapur, A.P, India
mail_hareeshk@yahoo.com

[2] Professor and Chairman, Department of CS, Mangalore University, Mangalore, India.
ylm321@yahoo.co.in

ABSTRACT

As the various architectures and protocol have been implemented a true VoD system has great demand in the global users. The traditional VoD system does not provide the needs and demands of the global users. The major problem in the traditional VoD system is serving of video stream among clients is duplicated and streamed to the different clients, which consumes more server bandwidth and the client uplink bandwidth is not utilized and the performance of the system degrades. Our objective in this paper is to send one server stream sufficient to serve the many clients without duplicating the server stream. Hence we have proposed a protocol and algorithm that chains the proxy servers and subscribed clients utilize client's uplink bandwidth such that the load on the server is reduced. We have also proved that less rejection ratio of the clients and better utilization of the buffer and bandwidth for the entire VoD system.

KEYWORDS

Video-on-Demand (VoD), video, Chaining, Segment, Buffer, Bandwidth.

1. INTRODUCTION

The internet is a global system, which serves the billions of global users today across worldwide. The users are access the vast range of information resources, services and new features from it. The most traditional communications such as telephone, music, film and television are redefined by the internet and provide new features and services include voice over Internet protocol (voIP) and IPTV. IPTV[1][2] is a new system provides lot of features and multimedia services likes Live TV, Video on demand(VoD) and interactive TV etc., reliably and secure in quality content distribution over IP based networks than traditional cable television fails to provide features such as high definition video quality and VCR functionalities Another important aspect that makes IPTV so popular and better than traditional cable television is Video on Demand (VoD) system. The applications involves, streaming video files from a multimedia server to the client systems that is large number of users on an individual basis [3][4][5]. This process involves streaming of videos requested by the clients in particular to them. In regular cable TV certain kind of customized viewing option is absent as the client received the requested video. The client requests for a scrupulous video on a multi media server and once the request is received by the multimedia server, which is process the request and served to the requested client. Through VoD, the video can be viewed as it is being downloaded from the server. In real work loads, high degrees of interactive operations were observed [6]. Along with this the VCR functionalities provide an extra edge over conventional television, letting the user decide when and how to watch the video.

The implementation of VoD requires a server with high bandwidth network framework which has sufficient storage capacity to store video files on the VoD network. One of the main requirements of a VoD network which uses IPTV generally uses High definition TV videos and hence servers need to have large storage capacities to store the video files. Using MPEG[7] standards the bandwidth required is 2133 Mbps. To keep up such a high bandwidth this requires a high bandwidth VoD network. This makes the service costly and not viable to implement from the client's side. To conquer these drawbacks the VoD network is implemented in three stages.

- Stage I has the media server with a fiber optic communication channel having in the range of 2 Gbps bandwidth.
- Stage II consists of cluster of proxy servers which are connected to the Media server at one side and group of clients on other side. The connection on the media server side is a fiber optic communication channel and that on the client side is a low bandwidth network that provides a bandwidth in the range of 2 Mbps.
- Stage III consists of a group of clients that are served by the proxy server and in turn served by the media server.

Initially, suppose if a client makes a request for a video to the media server then the server downloads the entire video data file to nearest proxy server of the requested client. This proxy server in turn requests to the server for scrupulous movie then server will streamed the movie data file to the proxy server and then streamed to the requested client.

When a client is requested a movie to the server that is previously being streamed to some client, again by the other client request for the same movie file thereby streaming the same movie to another clients that uses twice the bandwidth for the same movie file. The server bandwidth is of great importance in the VoD network and the streaming of the same movie twice which is not an optimal utilization of the server bandwidth. On the other hand client's side uplink bandwidth is not utilized. To overcome these draw backs we have provide a protocol for the better utilization of client's uplink bandwidth for streaming of the movies and hence reducing the load on the server which is our main objective of this paper.

2. RELATED WORK

The VOD streaming has attracted significant attention on many researchers from past decade; some of the previous works are as follows. There have been two kinds of works in this regard. They are: Building of Streaming servers Protocol and architecture related works.

Here, we are proposing an algorithm and protocol; our work will focus more on Protocol and Architecture related works. There have been quite a few innovative techniques and protocols proposed in this field. In prefix caching [8] concentrates more on the storage problems related to proxy servers, but in our work, when a client makes a request to the proxy server. It checks in the current movie list of the proxy server. If the movie is available then it streams from the proxy server to the requested client, otherwise the movie is downloaded from the media server to the proxy server and then the movie is streamed to the requested client. In the present days where memory is cheap and easily available the impact of prefix caching is very less and the proxy server which are used having more storage space.

In optimized content distributed delivery [9] method a single multimedia file is split into a several parts and transmits in parts from the various different servers to the requested clients. Effectively this is like many clients are served by the various different servers. The drawback of these systems is that, the server bandwidth is costlier and very expensive to maintain the multiple servers In our work, the proxy server has classes of clients based on the client's resources and scarce uplink and downlink bandwidth with limited buffer. These proxy servers are connected to the main media server through T1/E1 link. Each client acts as a server and has sufficient resource to store the video stream as well as forward the buffer to the nearer clients using its uplink bandwidth. By this way, more number of clients is chained for single server stream and which is less expensive inturn reduce the server bandwidth.

In hybrid model [10] uses many clients called peers-lend or supply their resources to the network and serve these resources to their peers. This model efficiently uses all the network resources and provides very good services but the major drawback is that, the protocol requires the services among all the peers in a network and hence the service becomes not reliable and the model is too difficult to implement In our work, the optimal utilization of the server bandwidth is proposed using protocol and implemented using Chaining algorithm. Hence the protocol requires services of each and every peer and the service become more reliable.

To further improve the proxy or server load resources, in [11] proposed a client-side cache scheme called earthworm. In this scheme, the client not only playbacks the movie but also forwards the stream to another client with adequate buffering delay. This is called as basic chaining. Further, this can be combined with batching period, which is called as extended chaining [12] in both the schemes, it uses backward buffering. Therefore, when client buffer overflows, then the chain has to be broken and a new server stream is required. To address this problem, an adaptive chaining [13] scheme is used. This achieves double the performance of the above-mentioned chaining schemes, in which both forward and backward client buffers are used for streaming the movies. However, the combined length of the forward and backward buffers of the client is less than length of the video segment. Hence, to extend the chaining as much as possible, an optimal chaining [14] is proposed in which the buffer of the remaining clients is exploited. This is achieved by patching stream during playback of the first few video segments. The major drawback of this scheme is over-utilization of the bandwidth without client request. The buffer of the neighboring clients is exploited by assigning the segments. Due to this, tremendous transfer of same data occurs in the network, which will degrade the overall performance of the network. In our architecture, we combine the non-scaled proxy server based architecture and decentralized P2P overlay network which reduces the overall cost of the system and improve the system performance.

3. SYSTEM ARCHITECTURE

The system network is the combination of proxy based architecture and peer to peer network based system as shown in Figure 1.The network architecture consists of a media Server, a cluster or region we find Proxy servers and many clients are connected through proxy server to the media server. Each Proxy server in a cluster act as a regional server which serves number of clients that is available in a particular region. One of the aspects of having proxy servers is that the probability of more than one client requesting the same movie in a given region.

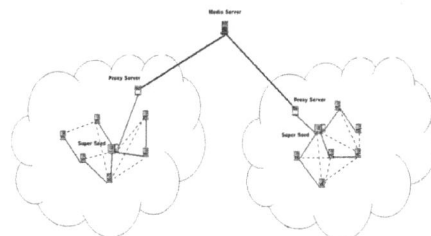

Figure 1: Network architecture

Since, if the movie is more popular, the probability of clients requesting the same movie will increase. Hence, quite obvious that number of clients requesting for popular movie will be greater in number. Thus if the Server satisfies each and every request for the same movie, the server bandwidth is not optimally utilized. In order to avoid this problem, these popular movies will be stored in the regional Proxy server buffer and streamed as and when the client requests for it. Since the popularity of each movie is depends on the region based, each proxy server buffer contains the movies which are popular in a given region and the movies can be streamed directly to the requested clients. The popularity is based on the number of hits that occur for a particular movie. The movies are stored dynamically in the proxy server based on the popularity, that if the popularity of a particular movie will decrease, the movie is automatically deleted from the proxy buffer and in turn provide to some other more popular movie. The above mentioned parameters comprise the essential requirements for the proposed protocol.

Whenever a movie is requested by a client, instead of being served by the server or the proxy server, the request is redirected to another client who already contains the same video blocks in its buffer; this process is called Chaining. This reduces the load on the proxy and the media server. Hence the performance of the system increases and better utilization of the server uplink bandwidth is implemented using Chaining algorithm.

3.1 Proposed Protocol

In this section describes how the protocol and algorithm is implemented. Whenever a client requests for a new Movie Id to the request handler, it checks the user details, and passes that request to media server. Media server checks existence of requested movie in its buffer. If the requested movie exists then the media server communicates with chaining handler to precede the serving operation. In the absence of movie in its buffer, chaining handler initiates proxy navigation module or cache module in order to get requested movie. The chaining handler checks active and passive list of clients and intimates client details that stores same movie (within its region) to requested client, and thereby initiates chaining operation. In case of absence of requested movie in both active and passive list of clients, proxy server starts streaming operation of requested movie to requesting client.

Proxy server executes given algorithm, upon receiving a movie request from a client. First proxy checks requested movie in its active list of clients. Then proxy server checks for requested movie in its passive list of clients In case of absence of requested movie file, proxy server initiates searching operation in its nearest proxies The proxy server executes pop check procedure to check popularity of given movie falls below the threshold values. Proxy server compares popularity, timestamp values to find out the existence of given movie in client video buffer. If popularity of a movie increases, automatically the life time of that movie in client video buffer also increases. In the absence of movie in client video buffer, proxy server starts streaming operation of movie. A client system, after receiving a chaining request from proxy server, executes given client chaining algorithm, as specified below.

3.2 Algorithm

The algorithms that are required to implement the above protocol are as follows:

CHAINING ALGORITHM

Nomenclature:
$C=\{C_1, C_2, C_3, C_4.....C_m\}$, Active clients $P=\{P_1, P_2, P_3, P_4........P_n\}$, Active Proxies
$C_i \rightarrow$ Client address
$P_i \rightarrow$ Proxy address
$B_i \rightarrow$ Buffer size used in client C_i

$b= \{(\mu_1, \vartheta_1), (\mu_2, \vartheta_2)... (\mu_i, \vartheta_i)\}$, b is a list containing μ_i and ϑ_i

Where μ_i is Movie Id, ϑ_i is popularity.

$\mathbf{D}\{i, j\} \rightarrow$ Distance between Client C_i and C_j

$\mu \rightarrow$ Movie Id (Request Id)

$\eta_i \rightarrow$ Uplink of client C_i

$\mathbf{R} = \{ r_1, r_2, r_3......r_k\}$, A list containing current streaming request.

$Bi \rightarrow$ Proxy P_i's current Bandwidth

$\beta \rightarrow$ Maximum allowable buffer size of proxy P $S_r \rightarrow$ Requested   segment

$Si \rightarrow$ Segment number i

$\mathbf{Q} \rightarrow$ A circular queue in client holding received segments

Proxy Server

Upon receiving request from client C_i

1 if μ in R then

2 find j such that $(cost*i+ \rightarrow \mathbf{D}\{i, j\}/ \eta_i) < \mathbf{D}\{i, p\} / B_i$, where $r_j \rightarrow \mu$

3 if $\exists$ some j such that $C_j \in \mathbf{C}$

4   chain client C_i to C_j

5        $\vartheta_i \rightarrow \vartheta_i+1$, where $ri \rightarrow \mu$

6 else

7   if μ in b then

8   initiate new stream r_i to client C_i,

        $\mathbf{R}=\mathbf{R}\ U\ \{r_i\}$

9        $\vartheta_i \rightarrow \vartheta_i+1$, where $r_i \rightarrow \mu$

10 else

11   send request $r_k \rightarrow \mu$ to server

12   receive segments from server

13 if maxsize$\{b\} > \beta$

14   replace (μ_k, ϑ_k) in b with (μ_l, ϑ_l)

        Where $\vartheta_k < \vartheta_j\ \forall\ j$

Client

Send a request r_i to proxy P_j

Upon receiving SIG_READY from P_j

1 Receive segment S_O from P_i, $Q= Q\ U\ \{S_O\}$

2 cptr=0,i=0

3 if $B_i < size(S_i)$

4        $Q= Q - \{S_{in}\}$

5        $Q= Q\ U\ \{S_i\}$

6   cptr $\rightarrow$ cptr+1

        Upon receiving SIG_CHAINTO

7 If S_r in Q

8   stream the segments S_r to S_n

9 else

10 reject

If both the clients belong to same region, and requested movie exists in client video buffer then chaining operation starts. If both the clients belong to different region, chaining not possible message is given to proxy server. In the absence of requested movie file in client video buffer, chaining request is rejected. If the movie is not available in the proxy server or not available in any of the serving client, then the initial portion of the video blocks are directly streamed from the main server and the later portion of the video blocks are first downloaded and buffered in the proxy server and then streamed from the proxy server to the requested client. Perceptive behind this idea is all the customers and services providers are profited by better utilization uplink bandwidth and the system resources.

4. SIMULATION AND RESULTS

The following are the assumptions made in the simulation model: It consists of a single server, 5 proxy servers and 80 clients. The server had 20 movies each of size 4.9 MB. The proxy buffer was large enough to hold 50% of the data on the server that is effectively10 movies. The server to proxy bandwidth is 2 Gbps and the proxy to client bandwidth is 2Mbps. The uplink bandwidth of the client is 256kbps.

The graph shown in Figure 3 is the combined graph of "Number of clients' v/s Time" and the "Number of streams v/s Time".

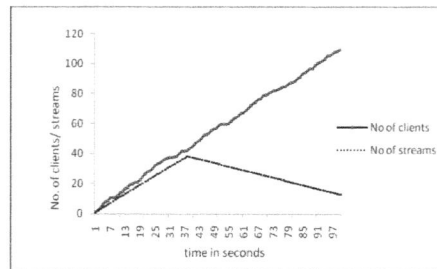

Figure 3: Number of clients and Number of Streams v/s time

We can observe that as the time increases the number of clients will also increase. But as the number clients increases we can observe that the number of streams increases initially and then decreases gradually. This is because the proxy or the server serves the clients only till all the popular movies have been cached in at least one place. Later on the movies are chained from these places.

The graph in Figure 4 shows the number of requests /streams with respect to time. In this case the requested movie is very popular and hence with time the number of requests for the movie is increased. We can observe that as the number of requests increases, the number of active chains also increases.

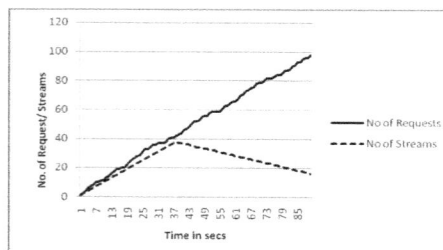

Figure 4: Number of streams v/s time

Initially the movie is present in very few clients, who are scattered in different locations. Hence the more number of active chains have to be created in order to serve the requests. But as the requests are served, the number of clients that have the movie will grow. Once more clients possess the movie, the number of active chains gradually decreases. The same as with the streams allocated for the movie in the server as shown in the graph. But the number of streams is always less than the number of active chains (as seen in the graph, the number of streams is always below the number of active chains) because of the chaining. The chaining reduces the number of streams from the server and hence the server bandwidth is conserved.

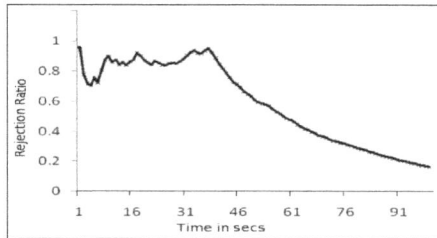

Figure 5: Rejection ratio with time

The graph in Figure 5 shows Rejection Ratio with respect to time. The rejection ratio is defined as the ratio of the number of rejected requests to the number of active chains. It represents the fact that the client requests are rejected even though there is a possibility of chaining. This happens because even though the chaining is possible the cost of chaining exceeds the actual cost of streaming. The ratio is around 0.9 initially since the number of clients possessing the movie is small and scattered. But as the number of clients that possess the movie increase the rejection ratio comes rapidly as shown in the graph. We can observe that the bandwidth initially increases with time and then peaks at a point and then reduces gradually to some constant value.

Initially the numbers of requests that arrive have to be served by the server itself because the number of clients that possess the movie is small, as shown in the previous graph of the rejection ratio.

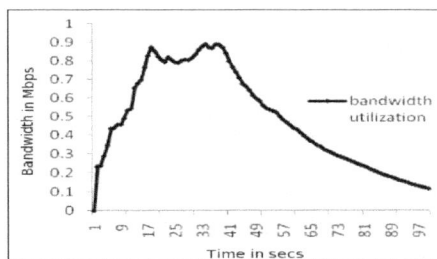

Figure 6: Bandwidth utilization with time

But once the chaining process is activated the bandwidth used of the server is reduced and hence the burden on the server is reduced to a great extent.

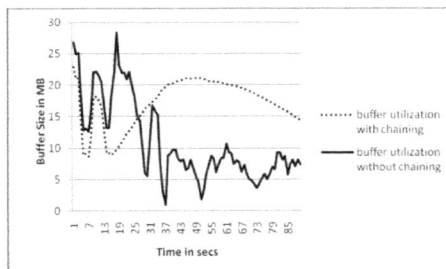

Figure 7: Buffer utilization

The buffer utilization graph shown in Figure 7 shows the client buffer utilized. Without chaining we can observe that the buffer utilization is erratic and decreases drastically with time. But with chaining we can observe that the buffer utilization is initially high and then reaches the peak and then falls down gradually. The peak represents the peak of the chaining process. After the completion of the chaining and the after the movie has been fully watched by the client the buffer utilization falls; this is the dip in the graph.

5. CONCLUSION

The Chaining reduces the overall system bandwidth and buffer utilization. In this paper we studied the existing architecture such as content delivery network; proxy based approach and peer to peer overlay networks, which suffers from the low performance by over utilization of bandwidth and buffer for the same movie. To improve the performance, we studied basic, extended, adaptive and optimal chaining schemes. The drawback of these schemes is the utilization of bandwidth is not efficient hence in this paper we have proposed a protocol and algorithm for better streaming of video streams through chaining process to the requested client using the client uplink bandwidth. Through simulation we have proved the advantages of using this protocol in the real world environment. The protocol uses the clients' uplink bandwidth, the server bandwidth is optimized and also the client's uplink bandwidth that was previously under used is well utilized.

We intend to build on the current system and make some improvements to make the rejection ratio even lesser as future enhancements. This will include procedures that will consider the popularity and the cost of streaming, in order to dynamically maintain the movies in the proxy and the servers. The algorithms used in this protocol are not equipped to handle fault tolerance. In case the connection between any two clients, which are involved in a chain, breaks, our algorithm does not provide a mechanism to overcome the drawback.

REFERENCES

[1]   B. Alfonsi." *I want my IPTV: Internet Protocol television* "predicted a winner, 2005.

[2]   M. Cha, P. Rodriguez, J. Crowcroft, S. Moon, and X. Amatrianin. *"Watching Television Over an IP Network"*. In Proceedings of ACM IMC, 2008.

[3]   J. Liu, B. Li, and Y.-Q. Zhang, *"Adaptive video multicast over the Internet,"* IEEE Multimedia, vol. 10, no. 1, pp. 22-33, Jan./Feb. 2003.

[4]   D.-E. Meddour, M. Mushtaq, and T. Ahmed. *"Open Issues in P2P Multimedia Streaming"*.MULTICOMM2006 Proceedings, pp.43-48, 2006.

[5]   X. Zhang, J. Liu, B. Li, and T.-S. Yum, *"CoolStreaming/DONet: A data-driven Overlay Network for Live Media Streaming"*. IEEE INFOCOM'05, Miami, FL, USA, pp.2102-2011,March, 2005.

[6]   RochaM,MaiaM, Cunha I, Almeida J, Campos S *"Scalable media streaming to interactive users"*. Proc. ACM Multimedia, pp 966–975, November 2005

[7]   ISOIIEC JTCI/SC29 WG111602, Information Technology –*"Generic Coding of MovingPictures and Associated Audio"*, Recommendation H.262, ISOIIEC 13818-2 CommitteeDraft, November 4, 1993.

[8]   Bing Wang, Subhabrata Sen, Micah Adler, and Don Towsley, *"Optimal Proxy Cache Allocation for Efficient Streaming Media Distribution"*, IEEE Transactions on Multimedia, Vol. 6, No. 2, April2004.

[9]   Gerassimcs Barlas and Bharadwaj Veeravalli, *"Optimized Distributed Delivery of Continuous-Media Documents over Unreliable Communication Links"*, IEEE Transactions on Parallel and Distributed Systems, Vol. 16, No. 10, October 2005

[10]   Mohamed Hafeeda and Bharat Bhargava," *A Hybrid Architecture for Cost Effective On-Demand Media Streaming*", FTDCS 2003.

[11]   Hua, K.A., Sheu, S. and Wang, J.Z. '*Earthworm: a network memory management technique for large-scale distributed multimedia applications*', Proc. IEEE INFOCOM'97, Vol. 3, Kobe,Japan, April, pp.990–997

[12]   Sheu, S., Hua, K.A. and Tavanapong, W. (1997) '*Chaining: a generalized batching technique for video-on-demand systems*', Proc. IEEE Int. Conf. Multimedia Computing and Systems,Ottawa, ON, Canada, pp.110–117.

[13]   Chen, J.K. and Wu, J.L.C. (1999) '*Adaptive chaining scheme for distributed VOD applications*',IEEE Trans.   Broadcast., Vol. 45, No. 2, pp.215–224.

[14]   Te-Chou Su, Shih-Yu Huang, Chen-Lung Chan, and Jia-Shung Wang,"*Optimal Chaining Scheme for Video-on-Demand Applications on Collaborative Networks*", IEEE Transactions on Multimedia, Vol. 7, No. 5, October 2005.

[15]   K. A. Hua, S. Sheu, and J. Z. Wang Earthworm: "*A network memory management technique for large-scale   distributed multimedia applications*", in Proc. IEEE Infocom'97, vol.3, Kobe, Japan, Apr. 1997.

[16]   J. K. Chen and J. L. C. Wu, "*Adaptive chaining scheme for distributed VOD applications*", IEEE Trans. Broadcast., vol. 45, no.2, pp. 215–224, Jun. 1999.

[17]   Te-Chou Su, Shih-Yu Huang, Chen Lung Chan, and Jia-Shung Wang,   "*OptimalChaining Schemefor Video-on-Demand Applications on Collaborative Networks*" ,IEEE Trans. multimedia, vol. 7, no. 5, october 2005.

[18]   Hyunjoo Kim and Heon Y. Yeom,*," P-chaining: a practical VoD service   scheme autonomic ally handling interactive operations, Multimedia Tools Appl*", Springer Science, Business Media, LLC2007.

[19]   Frederic Thouin and Mark Coates, McGill University, "*Video- on- Demand Networks: Design Approaches and Future Challenges*", IEEE Network, March/April 2007.

[20]   Schultz JJ, Znati T, "*An efficient scheme for chaining with client-centric buffer reservation for multi-media streaming*", Proc. of the 36th Annual Simulation Symposium (ANSS'03), pp 31–38,2003.

6

PRECISE POSITIONING SYSTEMS FOR VEHICULAR AD-HOC NETWORKS[1]

Samir A. Elsagheer Mohamed[1&3], A. Nasr[2&3], Gufran Ahmad Ansari[3]
[1]Electrical Engineering Department, Faculty of Engineering, South Valley University, Aswan, Egypt.
Emails: samhmd@qu.edu.sa; samirahmed@yahoo.com

[2]Radiation Engineering Dept., NCRRT, Atomic Energy Authority, Egypt.
Email: Ashraf.nasr@gmail.com

[3]Currentlly affiliated to the College of Computer, Qassim University, P.O.B 6688, Buryadah 51453, Qassim, KSA.
Email: ansariga1972@gmail.com

ABSTRACT

Vehicular Ad Hoc Networks (VANET) is a very promising research venue that can offers many useful and critical applications including the safety applications. Most of these applications require that each vehicle knows precisely its current position in real time. GPS is the most common positioning technique for VANET. However, it is not accurate. Moreover, the GPS signals cannot be received in the tunnels, undergrounds, or near tall buildings. Thus, no positioning service can be obtained in these locations. Even if the Deferential GPS (DGPS) can provide high accuracy, but still no GPS converge in these locations. In this paper, we provide positioning techniques for VANET that can provide accurate positioning service in the areas where GPS signals are hindered by the obstacles. Experimental results show significant improvement in the accuracy. This allows when combined with DGPS the continuity of a precise positioning service that can be used by most of the VANET applications.

KEYWORDS

Vehicular Ad-Hoc Networks; Positioning Systems; Neural Networks; Wireless Networks; Received Signal Stength.

1 INTRODUCTION

The number of vehicles increases dramatically day after day. This in turn gives rise to the urgent need for regulation of vehicular traffic and improvement of vehicle safety on highways and urban streets. Recently, there have been intensive efforts to networked-intelligent vehicles to yield the safety enhancement, and other potential economic benefits that can result from enabling both communication between vehicles and vehicular feedback to an ad hoc network. This gives birth to what so called Vehicular Ad hoc Network (VANET), by which, intelligent vehicles can communicate among themselves and with road-side infrastructure [1]. VANET can provide many useful services, such as collision avoidance, cooperative driving, automatic driving, navigation and probe vehicle data that increase vehicular safety and reduce traffic congestion, and offer access to the Internet and entertainment applications [1-4].

However, there are several challenges in VANET yet to be tackled before the full deployment of VANET technology in all over the world. Among these challenges is the knowledge of the accurate vehicle location that can lead to many safety enhancing applications. It is clear that a

---

[1] This work is supported by the Scientific Research Deanship, Qassim University, KSA.

precise positioning system that allows each vehicle to know its location in real time and continuously is a must for VANET to be operational.

There exist several positioning techniques that are suitable for many applications [1][5-8]. The most famous one is the GPS (by using a set of satellites that feeds information about the position of a GPS receiver). However, all the existing positioning techniques including the GPS have several drawbacks. Lack of accuracy of the resulting measurements is the most unacceptable disadvantage. For example, GPS devices can produce an error of up to 50 meters [7]. This accuracy may seem to be acceptable for several applications. On the other side, other applications like collision avoidance, automatic driving and lane tracking demand precise and accurate positioning information. Therefore, the existing positioning techniques including GPS are not suitable for this kind of applications in VANET. Recently an enhancement to the GPS referred as the Differential GPS (DGPS) consisting in installing expensive ground stations can improve the accuracy significantly. However, GPS and DGPS and the similar techniques do not work in tunnels, undergrounds, and in high dense building areas, because the signal cannot be received or received very weak.

Some VANET services must have continuous and accurate positioning information that cannot be interrupted, otherwise great damage or malfunction can take place: example of such application is the collision avoidance.

In this paper, we propose positioning techniques that can be used in conjunction with DGPS to provide a continuous and robust real time localization service with the following features: precise; not expensive (no extra hardware is required except those already needed by the VANET infrastructure); reliable; work in all areas and work in real time.

The rest of this paper is organized as follows: In Section 2, the related works and research efforts are given. Descriptions of the proposed techniques are given in Section 3. Experimentations and the obtained results are given in Section 4. Finally, the Conclusions and the future works are given in Section 5.

2 RELATED WORKS

There exits two categories of positioning systems: indoors techniques and outdoor techniques. Indoors techniques [5] are not suitable for VANET due to its cost and the limited distance they support. We focus here on the outdoor techniques. The most famous one which is always mentioned in the research papers is the GPS [7][9]. However, GPS has several drawbacks as discussed below:

- Its accuracy, the civilian GPS has a limited accuracy in the order of 50 meters in each direction. This is really unsuitable of the most of the applications of VANET, .e.g. collision avoidance, automatic driving (lane tracking), etc. A survey of the most famous GPS device vendors, the best announced accuracy is +/- 5 meters, which gives an error of about 10 meters in each direction. They say that this accuracy is obtained in 95% of all the readings. The other 5% of the readings, the accuracy may be very bad. This still too much far away of being acceptable in the critical applications requiring a precise positioning system.

- In some places (e.g. inside tunnels, undergrounds, etc.), the GPS has no coverage. This is because the GPS receivers calculate the position based on the signals received from 4 simultaneous satellites. In such places, the received signals are too weak or no signals can be received at all. Therefore, the GPS receivers cannot calculate the position.

- GPS signals sent from the satellites can be simulated. In other words, there exist GPS simulators (devices generating fake GPS signals). As mentioned before, the real GPS signals are relatively weak signals. GPS simulators produce relatively strong signals and therefore, the GPS receiver will use the fake signals and consider the real signals background noise. As a result, the receiver will calculate the position based on the fake signals. Thus, any jammer equipped by this kind of GPS simulator can inject erroneous data that leads to make the vehicle believe that it is in completely different position.

The existing outdoor positioning techniques are as follows.

- Signal-strength-based techniques. The receiver calculates an estimate of its location based on the received signal strength from several wireless access points. Lack of accuracy and the access points must be placed with care to cover the roads. Reported results based on this technique shows poor accuracy [1][11][20]. In this paper, we will show how to improve the accuracy based on this technique.

- Time-Of-Arrival (TOA)-Based techniques. It is based on the travelled distance from the base station to the receiver of a known signal. This is the solution adopted by GPS [9]. As we mentioned before, its practical measurements are not acceptable for VANET, as it lacks the accuracy. In addition, the Signal cannot be received in some places (tunnels, undergrounds, and near dense areas of building). TOA techniques require a perfect synchronization between the clocks of the base stations and the receivers. This cannot be guaranteed except by using atomic clocks which are very expensive. Otherwise the accuracy is compromised.

- Techniques based on the round trip time [9]. The receiver sends a small packet to the base station and wait for a reply from it. The elapsed time is proportional to the distance between it and the base station. By some calculations the receiver can know its location provided that it can talk to at least three base stations each one knows its accurate location. Again the distance between the base stations and the receiver must be large to obtain good results.

- There exist several other works that focus on VANET [8-15]. Most of these works uses the Received Signal Strength and provide a framework for using the GPS. However, the lack of accuracy in the obtained results makes the methods unsuitable for some of VANET applications.

3 DESCRIPTION OF THE PROPOSED POSITIONING TECHNIQUES

In this Section we will outline the proposed positioning techniques.

3.1   Introduction

As previously mentioned GPS, as a positioning system, is not suitable for most of the VANET services. The main raison for that is its accuracy. To solve this problem, DGPS system is used in the case if the accuracy is of great importance. DGPS can give a precision of several centimeters depending on the quality of the used components. The idea is to use some ground base stations that knows their accurate positions and broadcast the correcting information to the other GPS receivers in their covering areas. The receivers obtain from the satellites the normal GPS signals (which are not accurate) plus the correcting information from the ground base stations. As a result, the receivers can know in real time their accurate position. It should be noted that DGPS cannot operate if the normal GPS signals cannot be received from the satellites. The ground base stations provides only correcting information in the area. As a known fact, GPS signals cannot be received in the undergrounds roads, in the tunnels, under the bridges, near tall buildings or objects, or in areas having many buildings (like the case of city centers).

Road-Side Unites (RSUs) are essential components in VANET. The functions of the RSUs are to: a) disseminate the information to nearby vehicles; b) collect information to the central offices; and c) to provide Internet connection and entertainment to passengers. These RSUs are wireless Base Stations, a.k.a. Wireless Access Point (WAP) or AP for short. They are used to broadcast and receive information from the integrated VANET controller in each vehicle: in VANET terminologies, the integrated controller is named as On-board Unit (OBU). The OBU has an ad-hoc wireless communication capability. It incorporates also the unit that provides the vehicle by its current locations in real time; we refer to this unit as the OBU Localization Unit (OBLU).

Therefore, the natural question is: why not we use these RSUs to provide private, cheap, and accurate positioning system for the vehicles? This also can work in all the areas that have VANET infrastructure installed including tunnels, undergrounds, etc.

In order to have a positioning system suitable to the requirements of VANET services and application, we propose the following. The OBLU must contain a GPS receiver capable of the DGPS error correction operation. This part of the OBLU will provide the positioning information accurately in all the areas where the GPS signals from satellites are received. In addition, this part detects if there are correcting signals received from the DGPS ground bases stations or not.

The second part of the OBLU will be able to provide the positioning information in the following cases: in areas where the GPS signals cannot be received from the satellites; or if these signals are very weak; or in the areas where the correcting information from the DGPS ground base stations cannot be received or received too weak. This part of the OBLU will use VANET infrastructure and the next methodology to provide the positioning information.

We propose to use the Received Signal strength (RSS) from the RSU (which is an essential part of VANET infrastructure). The use the RSS as a positioning system for the indoor application is not new [1][5][11][14]. However, the reported results in the published works make this unsuitable to VANET (sometimes more than 50 meters as accuracy [11]). One contribution of this Paper is the new ways by which the RSS can be used to significantly improve the accuracy of the positioning (See Sections 44.2.2, 4.3.4 and 4.4 for more details about these improvements techniques).

3.2   Generic architecture of VANET possitioning system

Based upon the above introduction, we propose the following generic architecture of VANET positioning system that has to be followed in order to get a universal positioning system.

1) If the OBLU detects good GPS signals and at the same time it receives correcting information from the DGPS ground base station, then current location information will be obtained from the DGPS. We refer to it as P_{DGPS}

2) Otherwise, the following technique will be used to obtain the current location information. We refer to it as P_{RSS}

   a) On the road side, there must be several RSUs which are simply a wireless access points (WAP), that periodically broadcast their existence by sending beacons (small messages) that can be received by the OBLU in the vehicle in the coverage area.

   b) These RSU are planted on the side of the street. Each RSU is installed on a specific position. The RSU broadcasts its position within the beacons massage (the exact global position: the longitude, the latitude and the altitude). Each RSU must operate on a channel that does not interfere with the nearby RSU.

   c) The OBLU can measure the received signal strength intensity from all the RSUs in real time. The OBLU must hear from at least two RSUs, if the 2D location of the vehicle is desired. In the case of 3D positioning, at least three access points must be heard.

   d) If OBLU hears from many RSUs, this can improve the accuracy. For example, in order to obtain the 2D position of the vehicle on the street, only two RSU are need. But if it receives from three, we find three readings. By taking the average of the three values, the accuracy will be improved. Selecting which RSU to be used in the calculation is based on several conditions (See Section 4.5 for more detail).

e) The OBLU can determine its global positioning based on the RSS and the absolute position of each RSS. Details on this Step are given in Sections 4.3 and 4.4.

4 FIELD EXPERIMENTS TO MEASURE RSSs FROM WAP.

4.1 Experiments Description

Two field experiments are conducted in realistic conditions. VANET hardware is installed on a portion of a street. The RSUs are wireless access points (WAPs) operating on Wi-Fi 802.11g. The on-board unit is simply a Laptop having the software that can measure the RSS from the WAPs at any place in the covering area.

4.1.1 The first experiment

Let's first describe the environment and the precautions that we take for this experiment. We have measured the signal each 5m from the starting point 0 until the ending point at 200m. The elevation of the access points is 110 cm, above the ground. Distance between the line of access points and the parallel line of the test line equals 7m.   All the access points and mobile host operates on Channel 6. A diagram of the setup of this experiment is shown in Figure 1.

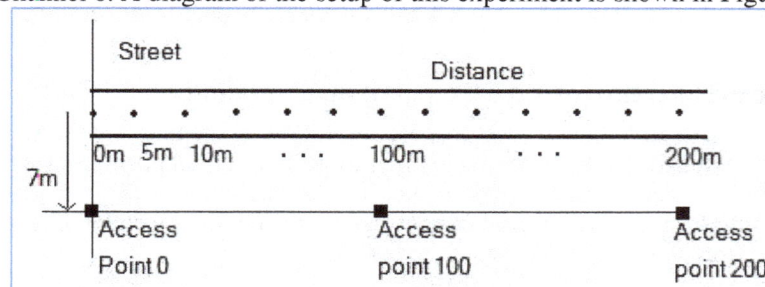

Figure 1. Experiment #1 Setup

4.1.2 The Second Experiment

In this experiment our goal is to solve a fatal problem discovered in Experiment #1. Mainly, to reduce the Wi-Fi channel interference and hence increase the accuracy of the RSSI, each WAP operates on a non-overlapping channel (Chanel #1, 7, and 13). All other experiment setup is the same as the previous one.

4.2 Obtained Results

In this Section, we will present the obtained results for the two experiments. In all the Figures in this Section, the x-axis shows the distance on the street (the longitude displacement of the vehicle). The y-axis shows the value of the received signal strength intensity (RSS) from the wireless access point(s) WAPs. Remember, the RSS is negative and the far we are from the WAP, the less the RSS.

4.2.1 Experiment #1 obtained results

Figure 2 shows the obtained RSS values from the three Access Points for the First Experiment (Experiment #1). From the shown data, we can see that the results are very bad, the data is very chaotic. It is expected that the RSS from any WAP decrease when we move far away of it (increasing the distance). However, for all the WAPs, this is not the case.

By inspecting the results, we can expect that no way to obtain accurate estimate of the position based on the RSSs. On other words, the accuracy will be very bad. That expectation is confirmed using the neural network as a learning tool. The accuracy is very bad, about 80 meters in any direction (for the lack of space, results are omitted).

A very interesting finding here is that this obtained result by the neural network (NN) for this experiment is almost the same as those found in the literature [9][11]. The published works for using the RSS as positioning technique suffer from the channel interference problem. Once we collected the results, we have thought in the problem. We, then, realized that all the WAPs operated on the same frequency channel. It is known that the Wi-Fi networks have 11-13 channels depending on the country. If all the WAPs use the same channel, that will cause a significant noise and interference. In such case, the estimation of the distance from the RSS is very difficult and hence the system gives inaccurate results. Thus, the interference problem has to be solved in order to improve the accuracy.

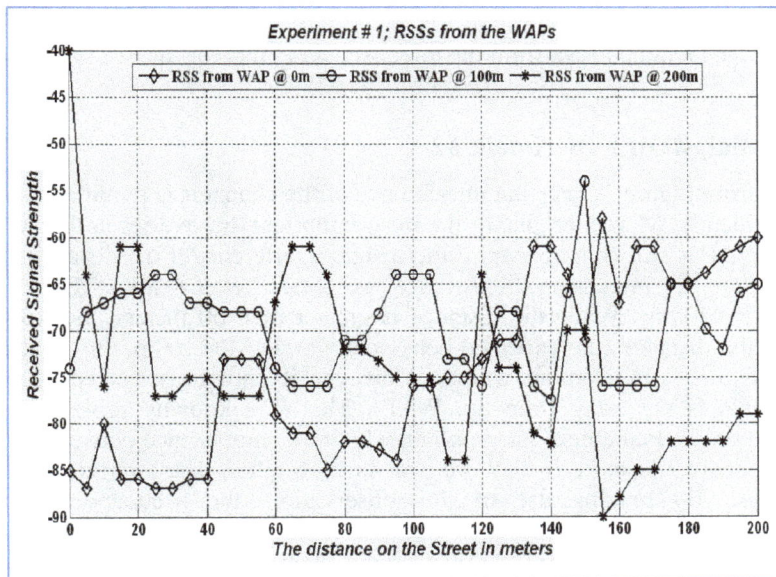

Figure 2. Obtained RSS from the three Access Points for the First Experiment (Experiment #1)

4.2.1.1 How to solve the interference problem?

It is also known that the channels do not overlap if they are more than 5 channels apart. For example, if one WAP use channel #1, the second uses channel #6 and the third uses channel #11, then there is not overlap and thus no noise nor interference that can degrade the results. That is confirmed using the obtained results.

4.2.2 Experiment #2 obtained results

Figure 3 depicts the obtained RSS values from the three Access Points for Experiment #2.

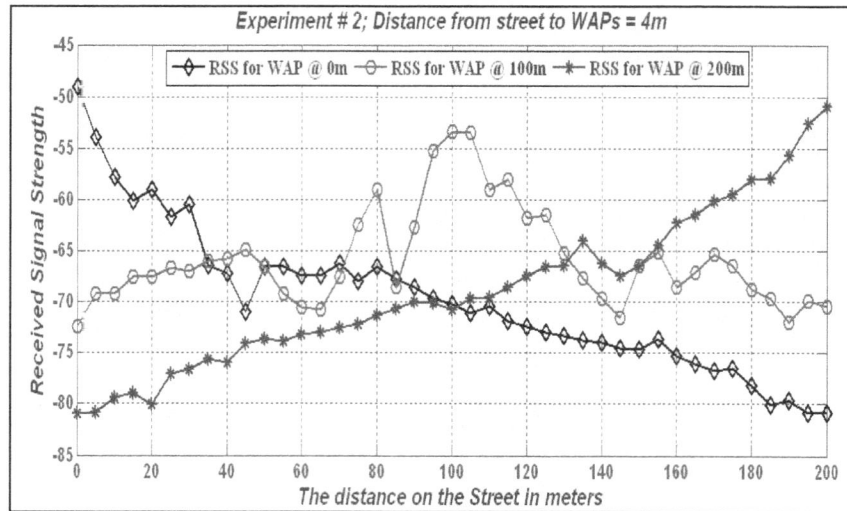

Figure 3. Obtained RSS from the three Access Points for the Second Experiment (Experiment #2).

4.2.3   Data Analysis of Experiment #2

As can be seen from Figure 3, once the interference of the channels is eliminated, the results are improved significantly. Moreover, one of the most important discoveries in this work is that the RSS from any WAP is still chaotic even if there is no interference if the distance to the WAP is less than 60 meters. By increasing the distance, we obtain better results. This can be seen for AP@0m and AP@200m. When the distance is greater than 60 meters, the RSS is inversely proportional to the distance (the expected behavior). For WAP@100m, this is also true. In the middle part of Figure 3, the data are chaotic. However, it starts to be correct at the boundaries (when the vehicle is far away from the WAP). This is a known problem in the wireless networks. It is referred as the near-far problem, where the interference between the emitter (the WAP) and the receiver is more if they are near to each other. The interference decays as the distance increase. By making use of this observation, the accuracy can be improved significantly. See Section 4.4.

In the next subsections we present the use of the neural networks and the curve fitting to estimate the position of the vehicle based on the received signal strengths and the use of the obtained experimental data.

4.3   A positioning system using the neural networks for VANET

In this Section, we present the use of the neural networks (NNs) as a learning tool to estimate the location of the vehicle on the road based on the RSS from the WAPs. The full theory and the description of the NN are beyond the scope of this paper.

4.3.1   Few words about the neural networks

The idea of the NN is inspired from the biological nervosystem. Similarly to the biological neuron, the artificial neural network learns by experience. In other words, a set of learning examples must be collected for the problem in hand. These samples are simply a set of the system inputs and their corresponding outputs.

Using training algorithms, the NN will learn the system behavior based on these examples. Once trained, the neural network will act as the system. For any new inputs, the NN will estimate the outputs based on the learning experience.

4.3.2   Description of the approach with the NN

We have used Matlab 2009a to do most of the calculations, and the Figures in this paper. In addition, there is a very nice, easy to use, and powerful toolbox in Matlab called the 'Neural Network toolbox'. The inputs to the NN are the RSS from the WAPs. The output is the location of the vehicle on the street or the required position. The NN needs to be trained, validated then tested. Thus, using Matlab, the whole learning example (the set of samples in the form RSS-Position pairs), is divided into three sets: one for training; the second for validation; and the third for testing. This is done automatically by Matlab. The ratios are 70% for training; 15% for validation; and 15% for testing. The training stops when the performance on the validation set increase. This is to avoid a known problem with the NN which is the overtraining.

For the NN, we use the Feedforward architecture that contains the input layer connected to the inputs; the output layer, which gives the estimate of the location based on the learning experience. In addition, there is the hidden layer. This layer can contain a given number of hidden neurons. Moreover, the random seed for the random generator which is used to assign the initial weights of the NN is very important. For the same samples, and the same NN, but different random seeds, the trained networks are completely different. Thus, we take this as a parameter for all the data.

 A Matlab script to automate most of the work is developed. Then the outputs can be saved and restored at any time to know the best NN that gives the best performance.

4.3.3   Working with data obtained in experiment #1

We have run the script on the data obtained from experiment #1. A total of 56 different NN architectures are created; trained; tested and validated. The performance measures are also calculated on both the testing and the whole samples.

Figure 4 shows the difference between the actual data (distance) and those estimated by the trained NN for the best one. As we can see from the Figure, error on the positioning may reach up to 60m. This is very high and not suitable at all for VANET. This bad result is that found on the literature, see for example [9]. Actually, people used the technique without taking into account the effect of the interference problem resulting from using the same channel for all the WAPs. This is why the published results for the mechanisms using the RSSs are not encouraging.

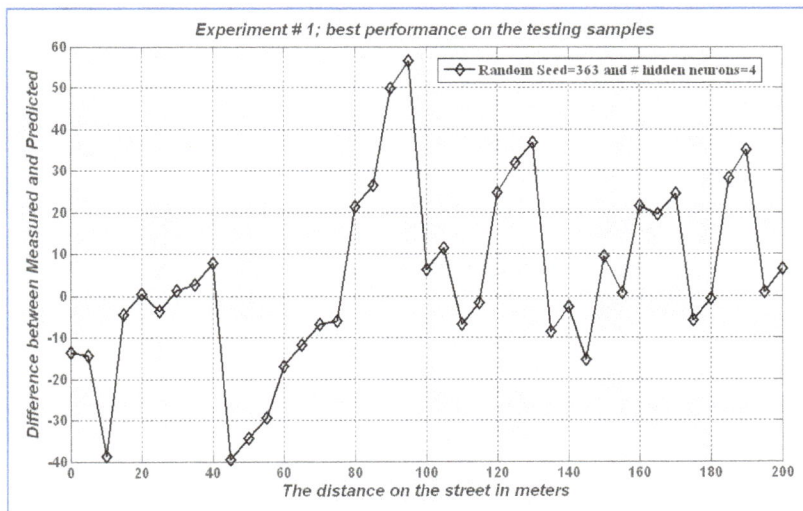

Figure 4. The performance of the best NN for experiment #1. The Figure shows the difference between the measured data experimentally and those calculated by the NN. The error can reach up to 60 meters. Correlation coefficients= 93 %

4.3.4   Working with experiment #2 data

Similar to what done in experiment #1, the whole work is repeated on the data for experiment #2. We have run the script on the data obtained from experiment #2. By using NNs with different hidden neurons and random seed generator: A total of 180 different NNs are trained, validated and tested. The performance measures are also calculated on both the testing and the whole samples. In Table 1, we show the obtained data sorted by the MSE on the whole samples and the max error obtained. In the table, we show also the standard deviation, the max error, the variance and the correlation for both the testing samples and the whole samples.

From this table, we can see that the best NN is obtained when the # of hidden neurons is 10 and for initial random seed =102. As expected the accuracy of this best one is very good compared to that of the first experiment. The MSE of the best NN is 6 against 172 for the first experiment. The standard deviation is about 3.2m, the max error is about 7.6 meters against 60 meters. Comparing these results to the GPS, we can see that we can reach accuracy better than the GPS. Thus, our technique can be used to complement the GPS in the areas where the GPS signal cannot be received or if it is very week.

Thus, the accuracy is improved significantly, by about a factor of 7. That is a huge improvement. This can also be seen from Figure 5 which depicts the difference between the actual data (distance) and those estimated by the trained NN.

4.3.5   Effect of removing the WAP@100m

We have removed the RSS from WAP@100m that we have thought may degrade the performance. The results show no further improvement (omitted for the sake of space). This is because three WAPs are better than two.

4.3.6   Using only one WAP but far away

We are seeking to improve the performance as much as possible. As described before, the near-far problem is a source of accuracy degradation. To examine its effect and the ability of the NN to learn, data from WAP @0m and WAP@200m only when the distance from each one is greater than 60m are used.

The obtained results also are not encouraging. Again, the NN cannot correctly learn from the few samples available for training. The next subsection presents another way to improve the performance, which is the curve fitting.

4.4   Positioning System Using Curve Fitting

In the previous Section, we have seen how to use the neural network for the positioning system. However, it is still in the range of 7 meters. We are seeking to improve this performance. We have explored the use of the curve fitting to obtain a polynomial equation that can be used to obtain the distance from the WAP given its RSS. Several types of curve fittings are evaluated using Matlab. The best result is given bellow. We aim to approximate the positioning problem according to the following equation. For the Linear model Polynomial of degree 4, we have the following equation:

$$Distance = p_1 R^4 + p_2 R^3 + p_3 R^2 + p_4 R + p_5,$$

where R is the received signal strength.

Table 1. Results of exp. #2. By using NNs with different hidden neurons and random seed generator: A total of 180 different NNs are trained, validated and tested; only the top 5 are shown

#	Hidden neurons	random Seed	Mean Square Error (MSE)		Max absolute Error		Standard Deviation		Variance		Correlation Coefficient	
			Testing Items	All Items	Testing Items	All Items	Testing Items	All Items	Testing Items	All Items	Testing Items	All Items
1	10	102	8.4	6.0	5.4	7.6	3.2	2.5	10.0	6.1	0.997	0.999
2	8	102	15.9	7.5	8.0	8.0	4.2	2.7	17.4	7.2	0.997	0.999
3	8	1	5.9	7.9	3.5	8.8	2.4	2.8	5.9	8.1	1.000	0.999
4	4	899	8.2	8.1	5.4	8.6	2.9	2.9	8.5	8.3	0.999	0.999
5	9	363	38.6	8.3	12.4	12.4	5.5	2.8	30.1	8.1	0.998	0.999

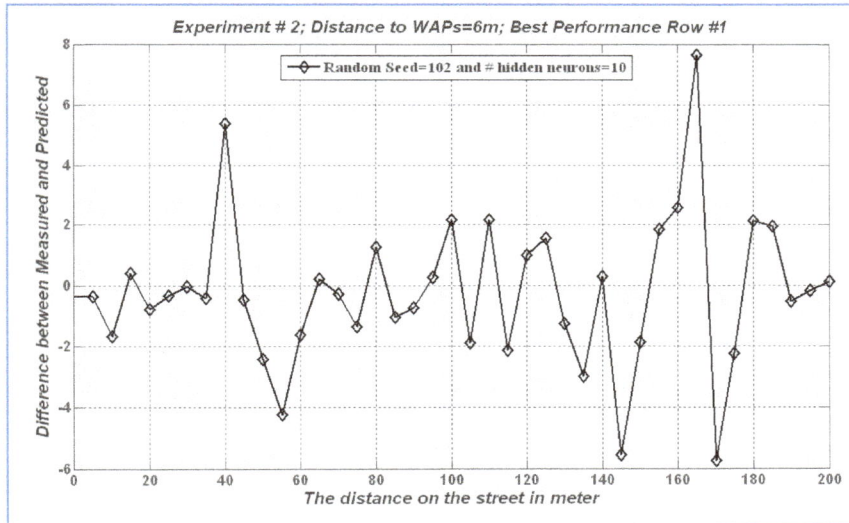

Figure 5. The performance of the best NN for experiment #2. The Figure shows the difference between the measured data experimentally and those calculated by the NN.

4.4.1   Starting from 60m

We have used the data from WAP@200m and by removing the data when the distance is less than 60m (the chaotic data). Again, this is to solve the near-far problem. Using Matlab "cftool" curve fitting tool, the estimated values of the Coefficients (with 95% confidence bounds) are as follows:

p_1= -0.005206, p_2= -1.553, p_3= -173.5, p_4= -8608, p_5= -1.601e+005.

The obtained Goodness of fit for the above data is: SSE: 422.7; R-square: 0.9917;   Adjusted R-square: 0.9903; and RMSE: 4.197. The curve fitting results are shown in Figure 6.

We have calculated the distance for the given RSSs in the range. In Figure 7, we draw the difference between the actual data (measured) and the calculated ones. One can see that using only one WAP, the accuracy is almost the same as using the NN with three WAP. That is a very important finding. Staring from 60m solved the problem of the near-far problem that the wireless networks are suffering from.

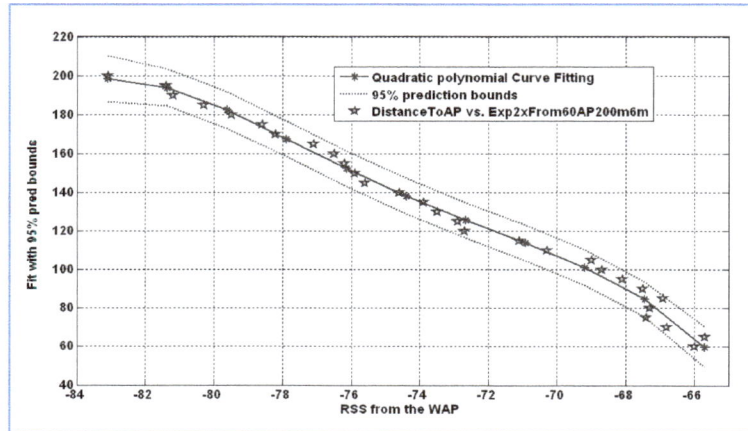

Figure 6. Curve Fitting results using cftool, when the Distance is greater than 60m

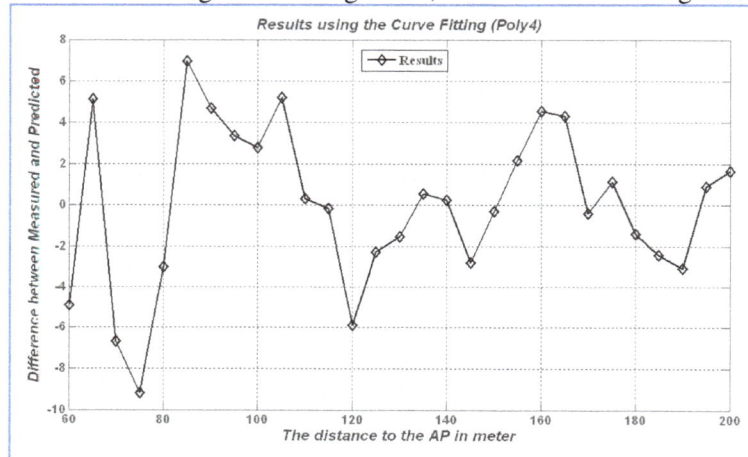

Figure 7 The performance of curve fitting for Exp # 2 by using only data from WAP@200m and excluding data before 60m. The Figure shows the difference between the measured data experimentally and those calculated by the NN.

4.4.2   Starting from 100m

From the previous section, we can see that the performance is bad at the beginning of the curve. However, it is getting better when the distance from the WAP increase. Thus, we redid the same experiment, but removing all the data when the distance is less than 100m. The results of the curve fitting are shown in Figure 8.

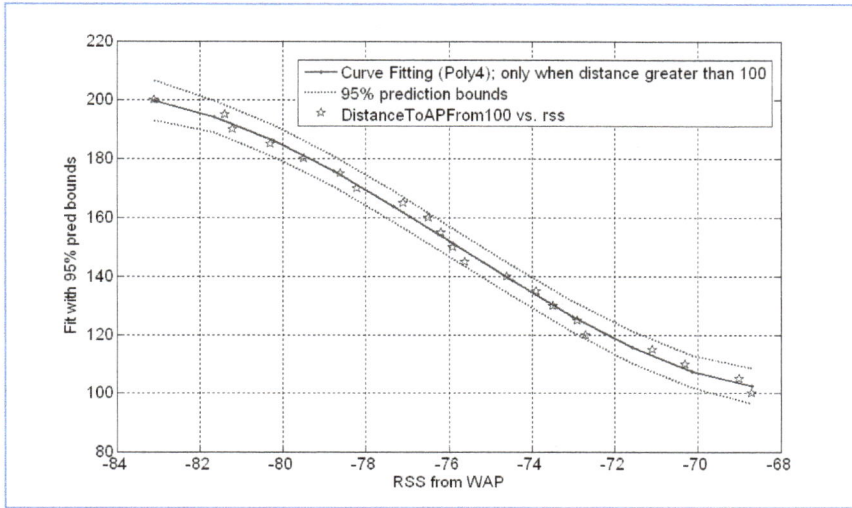

Figure 8. Curve Fitting results using cftool, when the Distance is greater than 100m

The coefficients of the polynomial equation are as follows: p_1=0.0004373, p_2=0.1736, p_3=24.4, p_4=1458, p_5 =3.171e+004. The obtained goodness of fit is: SSE=85.13, R-square=0.9956, Adjusted R-square=0.9945, RMSE= 2.307. Comparing these data with the ones in the previous experiment, we note a very significant improvement. This is also confirmed from Figure 9. We can see that the accuracy is improved (the max error is now 4 meters instead of 8 meters). This is by using only one WAP.

Again, if we inspect the figure carefully, we can see that the results when the distance is greater than 170m are within only 2 meters (A very accurate results). This result is much better that that obtained using the GPS. Moreover, it costs nothing (no extra hardware is needed, only the RSS from the RSUs), it can work in tunnels, undergrounds, and in the dense areas or near tall buildings where GPS cannot be used.

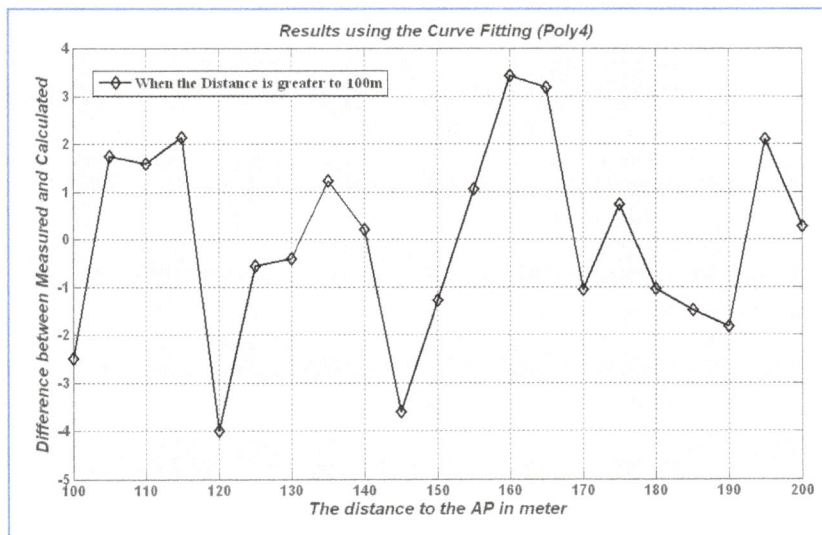

Figure 9. The performance of curve fitting for Exp #2 using only data from WAP@200m and excluding data before 100m. The Figure shows the difference between the measured data experimentally and those calculated by the NN.

4.5   The best approach for the positioning system

Based on all the results explained in the previous Subsections, we summarize here the best approach that can be used to build a system for an accurate positioning system using the RSSs from the RSUs for VANET. (See Section 3.2 for details on the generic system with conjunction of the DGPS.)

- The RSUs must be installed with 100m of minimum distance between each two.

- At any point on the road, the on-board unit must receive from at least two RSUs satisfying the next condition: these RSUs must be as far as possible (the less the RSS), and each operates on a different radio channel. This is to minimize the effect of the channel interference and the near-far problem.

- Using a simple triangulation technique, the 2D and the 3D coordinate point of the vehicle can be determined.

- If the on-board unit receives from more than two RSUs, this is very useful. That can be used to improve the performance. If there are extra RSUs, many locations are obtained for the same point due to the error of the positioning. Thus, the average must be taken by all the obtained locations to improve the accuracy.

- Each RSU knows its absolute location. Thus, the vehicles can determine its absolute location based on the estimated relative position and the absolute locations of the surrounding RSUs.

5 CONCLUSIONS AND FUTURE DIRECTIONS

In this paper, we have tackled the positioning problem for VANET applications. The proposed system is based on the use of the received signal strength (RSS) from several RSUs and the use of the DGPS.

For the use of the RSS, field experiments are conducted on realistic situations. The first one has a fatal error, which is the interference due to the use of the same wireless channel of all the WAPs. People who works on the same problem, uses this setup. Thus, their obtained results were not accurate. However, we have discovered the cause of this problem and how to solve it. This is manly by eliminating the effect of the channel interference and the near-far problems. In the second experiments, the results are very good compared to the first as the RSUs operated on different channels to reduce the interference. After analyzing the obtained results, we have proposed the use of the neural networks (NN) to learn the positioning problem from the obtained learning examples. The basic accuracy obtained is in the range of 60m for the first experiment (suffering from the interference). By using the data from the other experiments, the accuracy reached 8 meters (the maximum error). By using the curve fitting technique, and using only one RSU, we have done two experiments that show an obtained accuracy that can reach a maximum absolute error of only 2 meters. This can be obtained when the RSU is very far from the vehicle. The accuracy of determining the location can be less than 2m in the closed areas (tunnels, etc). We propose the use of DGPS in the areas where the GPS signal and DGPS correcting information can be obtained. In other areas, the proposed mechanism based on the RSS can be used.

In order to further improve the accuracy, we are currently exploring the use of nanotechnology device to determine the location precisely even in the areas where GPS signals cannot be received or very week.

REFERENCES

[1]   K. Saha and D. B. Johnson, "Modeling mobility for vehicular ad hoc networks," in *Proc. of the ACM Workshop on Vehicular Ad hoc Networks (VANET)*, October 2004, pp. 91–92.

[2]   R. Tatchikou, and S. Biswas, "Vehicle-to-vehicle packet forwarding protocols for cooperative collision avoidance", in Proc. of Globecom 2005, St. Louis, USA, Nov. 2005.

[3]   A.Nandan, S. Das, G. Pau, M.Y. Sanadidi, and M. Gerla, "Cooperative downloading in Vehicular Ad Hoc Networks", in Proc. of WONS 2005, St. Moritz, Switzerland, Jan. 2005.

[4]   S. Kato, S. Tsugawa, K. Tokuda, T. Matsui, and H. Fujii, "Vehicle control algorithms for cooperative driving with automated vehicles and intervehicle communications," *IEEE Trans. on Intelligent Transportation Systems*, vol. 3, no. 3, pp. 155–161, September 2002.

[5]   Hossain, A.K.M.; Hien Nguyen Van; Yunye Jin; Wee-Seng Soh; "Indoor Localization Using Multiple Wireless Technologies", In IEEE Internatonal Conference on Mobile Adhoc and Sensor Systems, 2007. MASS 2007. pp: 1 – 8.

[6]   Rahman, M.Z.; Kleeman, L.; Paired Measurement Localization: A Robust Approach for Wireless   Localization
        IEEE Transactions on Mobile Computing, Vol.: 8, Issue: 8, 2009, pp: 1087 – 1102

[7]   Yu, K.; Guo, Y.J.; Anchor Global Position Accuracy Enhancement Based on Data Fusion. IEEE Transactions   on
        Vehicular Technology, Vol.: 58, Issue: 3, 2009, pp.: 1616 – 1623

[8]   R. Singh, M. Guainazzo, and C. S. Regazzoni, "Location determination using WLAN in conjunction with GPS network," in *Proc. of the Vehicular Technology Conference (VTC-Spring)*, May 2004, pp. 2695–2699

[9]   M. Porretta, P. Nepa, G. Manara, and F. Gianne, "Location, Location, Location", IEEE Vehicular Technology Magazine, Vol. 3; pp. 20-29, June 2008

[10]   Ouyang, R. W.; Wong, A. K.-S.; Lea, C.-T.; Received Signal Strength-Based Wireless Localization via Semidefinite Programming: Noncooperative and Cooperative Schemes. IEEE Transactions on Vehicular Technology, Volume: 59, Issue:3, 2010, pp: 1307 – 1318

[11]   Bo-Chieh Liu; Ken-Huang Lin; SSSD-Based Mobile Positioning: On the Accuracy Improvement Issues in Distance and Location Estimations. IEEE Transactions on Vehicular Technology, Volume: 58, Issue: 3. 2009, Page(s): 1245 – 1254.

[12]   Bo-Chieh Liu; Lin, K.-H.; Jieh-Chian Wu; Analysis of hyperbolic and circular positioning algorithms using stationary signal-strength-difference measurements in wireless communications. IEEE Transactions on Vehicular Technology, Volume: 55, Issue: 2, 2006, pp: 499 – 509.

[13]   Mada, K.K.; Hsiao-Chun Wu; Iyengar, S.S.; Efficient and Robust EM Algorithm for Multiple Wideband Source Localization. IEEE Transactions on Vehicular Technology, Volume: 58, Issue: 6, 2009, pp: 3071 - 3075

[14]   Parker, R.; Valaee, S.; Vehicular Node Localization Using Received-Signal-Strength Indicator. IEEE Transactions on Vehicular Technology, Volume: 56, Issue: 6, Part: 1 2007, pp: 3371 - 3380

[15]   Misra, S.; Guoliang Xue; Bhardwaj, S.; Secure and Robust Localization in a Wireless Ad Hoc Environment. IEEE Transactions on Vehicular Technology, Volume: 58, Issue: 3 2009, pp: 1480 – 1489.

[16]   X. Chen, Y. Chen, and F. Rao, "An Efficient Spatial Publish/Subscribe System for Intelligent Location-Based Services", Proceedings of the 2nd International Workshop on Distributed Event-Based Systems (DEBS'03), June 2003

[17]   J. Hightower and G. Borriello, "Location Systems for Ubiquitous Computing," Computer, vol. 34, no. 8, pp. 57-66, IEEE Computer Society Press, August 2001

[18]   L. Wischof, A. Ebner, and H. Rohling, "Information dissemination in self-organizing intervehicle networks", IEEE Trans. intelligent transportation systems, vol. 6, no. 1, pp. 90-101, Mar. 2005.

[19]   C. Komar, and C. Ersoy, "Location Tracking and Location Based Service Using IEEE 802.11 WLAN Infrastructure," in *Proc. of the European Wireless Workshop*, February 2004.

[20]   S.A. Elsagheer Mohamed, "Why The Accuracy Of The Received Signal Strengths As A Positioning Technique Was Not Accurate?" International Journal of Wireless & Mobile Networks (IJWMN) Vol. 3, No. 3, June 2011. DOI : 10.5121/ijwmn.2011.3306

7

SECTOR BASED MULTICAST ROUTING ALGORITHM FOR MOBILE AD-HOC NETWORKS

Murali Parameswaran[1] and Chittaranjan Hota[2]

[1]Department of Computer Science & Information Systems, BITS-Pilani, Pilani, India
[2] Department of Computer Science & Information Systems, BITS-Pilani, Hyderabad, India

ABSTRACT

Multicast routing algorithms for mobile ad-hoc networks have been extensively researched in the recent past. In this paper, we present two algorithms for dealing with multicast routing problem using the notion of virtual forces. We look at the effective force exerted on a packet and determine whether a node could be considered as a Steiner node. The nodes' location information is used to generate virtual circuits corresponding to the multicast route. QoS parameters are taken into consideration in the form of virtual dampening force. The first algorithm produces relatively minimal multicast trees under the set of constraints. We improve upon the first algorithm and present a second algorithm that provides improvement in average residual energy in the network as well as effective cost per data packet transmitted. In this paper, the virtual-force technique has been applied for multicast routing for the first time in mobile ad-hoc networks.

KEYWORDS

Network Protocols, Wireless Network, Mobile Network, Routing Protocols, Multicasting,

1. INTRODUCTION

A mobile ad-hoc network (MANET) involves independent nodes that are rapidly deployed in an environment. These mobile nodes must make the correct routing decisions in the presence of node mobility, transient link failures and energy and other constraints. In the event of sending multiple copies of the same information to multiple nodes, use of multicast routing is the preferred option. Depending on how the distribution paths among multicast group members are created, multicast routing protocols can be classified into tree-based, mesh-based and hybrid multicast routing protocols [1]. Tree-based protocols can further be sub-divided into source-tree and core-tree protocols. In our algorithm, we take the middle path by letting the multicast tree be created by the initial packet in transit.

The creation of multicast tree can be equated with the Steiner tree problem. Steiner tree problem for a graph $G(V,E)$ is the problem of finding a tree spanning all nodes in Q, where $Q \subseteq V$, such that the length of the resultant tree is minimized [2],[3]. The set of nodes S, where $S \in V, S \cap Q = \varphi$, are known as Steiner nodes. A Steiner minimal tree constructed for a local sub-graph of a G need not be part of the minimal tree constructed for the entire graph [2]. Steiner tree problem is known to be NP-complete for planar graphs. Approximation algorithms like the one given in [4] have been suggested for the Steiner tree problem. A related notion is to classify a tree as relatively minimal Steiner tree. In a relatively minimal Steiner tree, the length of the tree is minimal for the specific set of nodes chosen to be part of the tree. The problem of multicasting reduces to finding a minimal Steiner tree in a graph. The task of a multicast routing problem is to determine the correct set of Steiner nodes S so that the tree can reach all terminal nodes in Q. For efficient communication, any multicast tree created must at least be relatively minimal. In our approach, we attempt to determine a relatively minimal Steiner tree with the help of virtual force approach.

The notion of virtual force had been explored in deployment problem in wireless sensor networks [5],[6], and in routing in MANETs [7]. In the virtual force approach (VFA), the participating nodes and/or the packet in transit are applied with some electric charge. The electrostatic forces are computed and the routing decision is made based on the magnitude and direction of the resultant force. Poduri, *et al*. [5] demonstrated the use of a combination of attractive and repulsive forces for solving sensor deployment problem in wireless sensor networks. Liu, *et al*. [7],[8] used the concept of virtual force in unicast routing for MANETs. In the routing algorithm for MANETs, the resultant force is used only to make a decision to move the packet forward.

Our main contribution in this paper is the use of virtual force in multicast routing for MANETs. The notion of virtual force is used to guide the packet towards the destinations. The effective force on the packet is a sum of contributing force values from the destinations, the source node and the packet itself. In addition to the forward guiding force, we use dampeners to limit choice of the next node in the path to accommodate QoS parameters. The combined result of these forces will let the packet know whether it is currently in a Steiner node. The resultant force will be towards the node that will be in the path to the majority of the destination nodes in one direction of a branch in the multicast tree.

In this paper, we have used the virtual force technique for multicasting. While attempting to use the technique for multicasting, we identified several new issues that were not present in unicast routing algorithms. The first algorithm proposed in this paper deals with directly imposing the virtual force technique for MANETs. We provide a refined algorithm that uses virtual force in a slightly different manner. We divide the region around the current node into sectors and perform a virtual-force based multicast routing on each of the sectors. Multicast routing using virtual force technique is attempted for the first time in this paper. As far as we know, no other paper has suggested the use of sectors of variable arc-lengths for performing multicast routing, in conjunction with the virtual force technique.

The remainder of the paper is organized as follows: section 2 describes the related work for our algorithm. Section 3 provides overall description of the technique, along with basic idea, assumptions used and the modelling details. Section 4 describes the first algorithm proposed, viz. multicast routing algorithm using virtual-force. Section 5 describes the details of the sector-based virtual force-based multicast routing algorithm. Section 6 contains the simulation results, while section 7 concludes the paper.

2. RELATED WORKS

Gilbert, *et al*. in [2] had put forth the criteria that the angle between out-going edges at a branch node for a Steiner node is 120°. A relatively minimal Steiner tree is a Steiner tree when this angle is greater than or equal to 120° for all internal nodes. In their paper, they have suggested an inductive property for constructing relatively minimal Steiner trees with the help of unit tension force from the group nodes. However, in a real-world network, we do not have control over the location of various nodes in the network. Thus, we should ideally be choosing branching nodes such that the angle between the out-going edges are exactly 120°, but still be able to handle cases where appropriate nodes are not available in those positions.

Mauve, *et al.* suggested a multicast routing algorithm using geometric information in [9]. Their greedy algorithm uses a heuristic dependent on the normalized number of next hop neighbours to determine whether a branching node in multicast tree has been reached. The parameter λ -used in their algorithm determines how late the split of the multicast forwarding will take place, with $\lambda = 0$ indicating an early split and $\lambda \approx 1$ for a very late split.

Fotopoulou, *et al.* [10] suggested the use of geo-circuits for unicast routing. They computed the unicast path from source to destination based on greedy position-based strategy and assigned a virtual circuit number to such paths. Once a geo-circuit was established, they re-used this geo-circuit to send future packets to the destination till the end of traffic or till the path is updated due to node mobility. The idea of using geographic virtual circuits can be used for multicasting as well, with a few minor revisions in the route caching system.

Liu, *et al.* suggested two routing algorithms, SWING [8] and SWING+ [7], for unicast routing in MANETs based on the notion of small world networks. These algorithms expanded the notion of virtual circuits as described by Fotopoulou, *et al.* [10] to include long virtual logical links to the surrounding neighbourhood of the current node with the help of virtual force. These two protocols were the first unicast routing protocols to use virtual force to compute the path towards destination. By computing the repulsive force exerted on the neighbours of the current node, the next hop neighbour was chosen as the node with the maximum repulsive force from the source node towards the destination node. They demonstrated that their approach works efficiently in a 3-D environment, as well. In this paper, we extend the notion of virtual-force suggested by Liu, *et al.* for multicast routing.

Rahman *et al.* [11] divided the region around the current node into quadrants, and considered four closest nodes in each quadrant for determining the next hop in the multicast tree. Their contention was that a split in multicast packet will occur when the destination nodes belong to multiple quadrants around the current node. They later expanded their algorithm in [12] used an intelligent energy matrix to increase the average life of the nodes in the network. Their algorithm provides a very reasonable length of the multicast tree. However, we found that by statically fixing the quadrants, their algorithm was generating slightly longer multicast trees.

The notion of virtual force as described in [7] cannot be used directly for the multicast routing problem. If we apply electrical charges in only the source node and the packet, as was described in Liu, *et al.* [7], we found that we could not really accommodate for multiple destinations typical in a multicasting environment. Initially, when we attempted to apply the electrical charges in destination nodes and the packet in transit, we found that the packet was moving towards a force-equilibrium position over and over again, increasing the likelihood of loops in the multicast tree. Also, supporting QoS parameters in a multicast route is slightly different than adding it in a simple virtual circuit because of the nature of path establishment. Instead of limiting the decision into fixed quadrants around the current node, as described in Rahman, *et al.* [12], we dynamically divide the region around the current node into α sectors based on the direction of the effective force due to the nodes in the multicast group. In this paper, we have devised simple mechanisms to adapt the concept of virtual circuits into the multicast environment.

3. MULTICAST ROUTING USING THE VIRTUAL FORCE TECHNIQUE

In this section, we start with a discussion on the assumptions and basic working of our multicast routing algorithms with the help of the virtual force technique. We then introduce the system model used.

3.1. Assumptions

We assume that location information is available in all nodes in the network and that all nodes in the network will become eventually aware of location and mobility information. Most off-the-shelf products currently come with GPS facility. Our algorithm assumes that there is an appropriate data structure that can store this location information. Such a data structure will allow look-up of node location from the node's address. As far as proper working of the algorithm is concerned, the location information in the immediate vicinity of the current node needs to be accurate. There can be slight imprecision in the location values of farther nodes, as the algorithm only computes the best likely members of the multicast route for the farther nodes. As the query for computing the multicast route reaches closer to the actual destination(s), the intermediate node can divert the query appropriately as it is assumed that the closer nodes will have more accurate information.

All mobile nodes are assumed to be having omni-directional antennas, as is the case with most of the off-the-shelf devices. As our algorithm relies on the neighbourhood information derived from the Hello protocol, all mobile nodes need not be in the same plane.

3.2. Basic Idea

In this algorithm, we assign positive charges to sender, the multicast destinations, and the packet. The packet is placed at a virtual point near the sender node, and will be guided by the effective repulsive force to move towards the nodes in the multicast group. All the other nodes are given a tentative charge based on QoS parameters.

When a packet reaches node u, we need to decide whether u is a branching node (a node at which the multicast tree branches) or not. If u is a branching node, we need to determine the number of branches to be taken, and the specific neighbours that will become part of the branch. If u is not a branching node, then we need to determine which of the neighbours in the forwarding direction has to be chosen for the next hop.

Let M be the set of nodes that are part of the multicast group. At the node u, we compute the effective force exerted on the packet due to nodes belonging to M, $\vec{F_M}$. We take the direction of $\hat{F}_M$ as the forward direction. We compute the effective force on each of the k neighbouring nodes v_1, v_2, ..., v_k in neighbourhood set of the current node, N_u, by taking into account the cumulative effect of dampening force($\vec{E}_{v_i}$), u and M. By looking at the effective force on the neighbours, we determine the p neighbours w_1, w_2, ..., w_p which form part of the multicast route, and forward the query to them.

3.3. System Model

Liu, *et al.* [7] defined Equation 1 for computing the virtual force the current node v to the destination v_d, where d_{max} is the maximum distance measure, and $d(v,v_d)$ is a measure of distance between v and v_d. This equation is sufficient for dealing with a unicast transmission, and for the computation of simple, point-to-point force between any two nodes. However, for a multicast transmission, we need to expand the notion to incorporate all the multicast group nodes.

$$F_{dest}(v,v_d) = d_{max} - d(v,v_d) \tag{1}$$

Alternatively, a definition of destination force as given in Equation 2 can also be used, where Q is a large constant charge value assigned for ease of computation. This is a minor modification of Equation 4 in [7].

$$F_{dest}(v, v_d) = \frac{2 \times Q}{d(v, v_d)}, v \neq v_d \tag{2}$$

At any node in the network, we compute the effective force on the node u due to the current set of destinations, M, as shown in Equation 3. Here, $\overrightarrow{F_{u,M}}$ is effective force on the node u exerted by all the nodes in the set of current multicast destination, M.

$$\overrightarrow{F_{u,M}} = \sum_{d \in M} \overrightarrow{F_{u,d}} \tag{3}$$

Apart from this force acting on the packet, there is a force component from the source node s and a dampening force caused by other parameters determining the choice of the next node. The effective force on the packet p at node u is given in Equation 4.

$$\overrightarrow{F_p} = \overrightarrow{F_{s,u}} + \overrightarrow{F_{u,M}} - \overrightarrow{E_u} \tag{4}$$

The dampening force $\overrightarrow{E_u}$ produced by the node u depends on the various QoS parameters that are considered. If k parameters are considered, the weightage of the ith parameter is α_i, the value of the parameter i in the node u is δ_i^u, the requested value for the QoS parameter is δ_{req} and the maximum value allowed for that parameter is λ_{max}, then the value of the dampening force at node u is given by Equation 5. Here, for each parameter i, $0 \leq \beta_i \leq 1$ and $\sum \beta_i = 1$.

$$\overrightarrow{E_u} = \left(\sum_i \frac{\alpha_i \cdot \left(\delta_{req} - \delta_i^u \right)}{\lambda_{max}} \right) \cdot \hat{F}_{s,u} \tag{5}$$

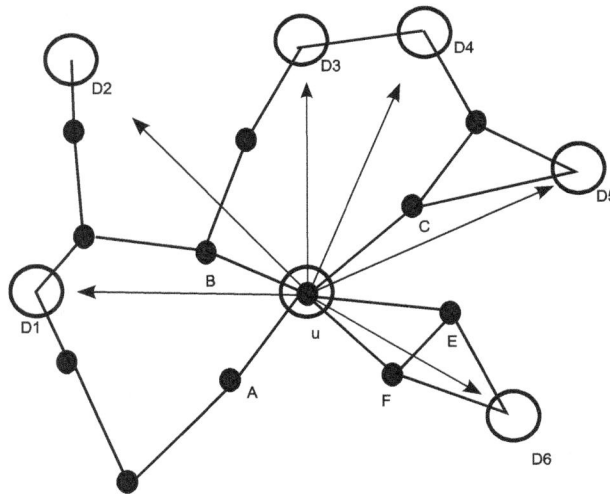

Figure 1. A sample multi-cast split node.

4. MULTICAST ROUTING ALGORITHM USING VIRTUAL-FORCE

In this section, we discuss the design of the Multicast Routing Algorithm using Virtual-force (MRAV) and its pseudo-code.

4.1. A sample working of MRAV

Figure 1 shows the node u that needs to multicast to six members of the multicast group M={D1,D2,D3,D4,D5,D6} represented as circles. Other members of the MANET are indicated as black bubbles. Viable communication links are indicated using lines. The node u has five neighbouring nodes, namely A, B, C, E and F.

Let us consider the example of a multicast from node u to the set of destination nodes depicted in Figure 1. The packet at node u experiences virtual force as a result of interactions with all the six destination nodes. The direction of the force is indicated with the help of arrows in the figure. We first compute the effective force on the packet that is currently at node u to each of the destination nodes. The arrows in the figure indicate the direction of the force vectors due to the six destinations. By computing the effective force on the node, we can compute the general forward direction of the node.

The Figure 2(a) shows the various force values on the current node, u, of the network shown in Figure 1. Here the resultant force at the node u is computed as vector addition of all forces at the point corresponding to node u. For the purpose of illustration, let us assume that the forward direction, which is the direction of the resultant force, is towards D3. Figure 2(b) shows the cumulative sum of forces along the axes at node u. Since there are destination forces affecting more than two directions in Figure 2(b), it can be inferred that the node u is a branching node, also termed as a split node. It now has to decide which among the nodes in its neighbourhood set (N_u) will form part of the multicast route.

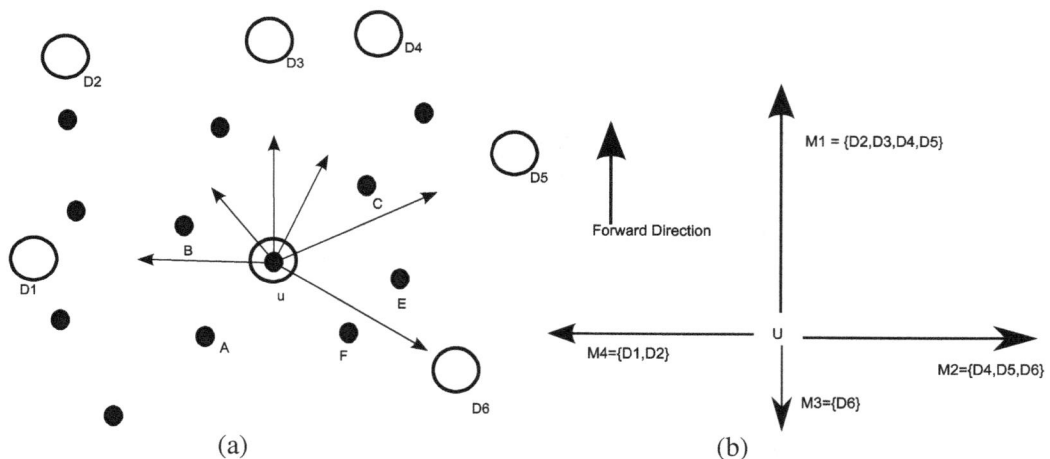

Figure 2(a). The virtual forces exerted by the destinations on node u. (b) Effecting force along axes.

Let us have four subsets of the destination set M, named as M_1, M_2, M_3 and M_4, indicating the set of destination nodes that had contributed to the force vector upwards, rightwards, downwards and leftwards respectively. In Figure 2(b), the elements belonging to the subsets are indicated along with the force component to which they contribute. It can be seen that D6 belongs to the destination subset M_3, M_2 contains {D4, D5} and M_4 contains {D1, D2}. M1 contains D2, D3, D4, and D5.

From Figure 2(b), it can be inferred that only the destination D6 is contributing to the downward direction, which is the direction opposite to the direction of the effective force at u. Also, in a clockwise direction starting from the left-side, it can be observed that the angle between D1 and D6 is greater than 180°. Because of the factor that it is the only node in subset M_3 as well as being more than 120°, D6 is considered separate from the remaining set of destination nodes. To reach D6, the relevant neighbours of u are the nodes E and F. If E is chosen as a multicast forwarding node for u, then both destination D5 and D6 can be served through it. In our algorithm, all nodes that are within ± 60° from the direction of the chosen force vector are considered to be part of the same general path towards multicast set. The Query message is forwarded to node E, with destination set as {D5, D6}

When the multicast route query reaches the node E, it identifies that D6 is its neighbour. So the node E directly communicates the query message to D6. To communicate to D5, however, there is no path in the general direction towards it. Thus choice of E as the next node neighbour for communicating to D5 is a wrong one.

To circumvent this issue of wrongly identifying a neighbour to forward the multicast query, we first evaluate all possibilities first, before deciding on the list of most suitable forwarding nodes. If the node still encounters a void, we choose perimeter routing as discussed in Greedy Perimeter Stateless Routing (GPSR)[13].

In this case, we first try to combine D5 and D6 as was discussed earlier, and then go on to determine the rest of the nodes to forward the query. The nodes D1 and D2 can be considered together, as they form part of the set M_4. The effective force that acts on the node B is the strongest among the neighbours of u. Hence, the forwarding node for these two multicast destinations will be B. The remaining nodes D3 and D4 can now be considered. The closest neighbour according to our force metric is node C. For the destination node D5, while comparing between the node C and the node E, it can be observed that C feels a stronger force than E. The destination D5 is hence removed from the *Query* message to node E and then attached to node C. The *Query* message is going to be forwarded to the nodes B, C and E with the destination sets marked as {D1, D2}, {D3, D4, D5} and {D6} respectively. Note that the node F can be used to replace E, as the effective force value on both E and F due to node D6 is the same.

4.2. The MRAV Algorithm

In this section, we discuss the multicast routing algorithm using virtual-force.

4.2.1. Data structures

Data structures relevant for the algorithm are mentioned in this sub section.

N_u : This is typically implemented as a linked list to store the list of neighbours of the current node. The list is updated with the help of responses from the periodic Hello packets transmitted as per the Hello Protocol.

P(M): This is the power set of M, such that all nodes within each of the subset created are reachable in increasing values of the angles. For example, in the sample network in Figure 1, P(M) may contain {D1, D2, D3}, but may not contain {D1, D3} as there exists a node D2 direction is in between D1 and D3.

Π : This is a priority queue that stores the subset of nodes with highest force value on the front of the queue. It is assumed that this queue also has separate lookup function implemented that can retrieve the force value for any subset given as input to the function. This can be achieved by including an auxiliary data structure in the form of a hash table along with Π.

v_s: The set of possible neighbours to choose from, while marking a split node. This list stores the next node as well as the list of destinations that are reachable from it.

4.2.2. Algorithm

The algorithm for sending a *Query* message is shown in Algorithm 1. Once the *Query* message has reached the destination or a split node, a virtual circuit linking the destination and the previous split node or the source node shall be constructed as the acknowledgement to the *Query* travels back to the source node.

Algorithm 1: Multicast routing Algorithm using Virtual-force

1:    **procedure** SendQuery(u,M)
2:    **if** $u \in M$ **then**   // The current node is itself a multicast destination
3:        $M \leftarrow M - \{u\}$
4:    **end if**
5:    **for each** $v \in N_u \wedge v \in M$ **do**   // A neighbour of the current node is a multicast destination
6:        Call $SendQuery(v, M)$
7:    **end for**
8:    **Let** M' = M
9:    **Repeat**
10:       **for each** s | s $\in$ P(M) **do**
11:           $\overrightarrow{F}_s \leftarrow calculateVirtualForce(u, s)$
12:           Add $\overrightarrow{F}_s, s$ to Π(M)
13:       **end for**
14:    **Let** M_s be front(Π(M))
15:       Compute $w \in N_u, s.t. \angle \overrightarrow{uw}, \hat{M}_s \leq 60^\circ \vee \angle \hat{M}_s, \overrightarrow{uw} \leq 60^\circ$
16:    **if** $M_s = M$ **then**   // all nodes are in the same general direction
17:           Call $SendQuery(w, M)$ and exit
18:    **end if**
19:       Add (w, M_s) to v_s
20:       Remove all entries from P(M) which contain the destination nodes M_s
21:       M' = M' - M_s
22: **while** M' $\neq \varphi$
23: **Mark** u as split node
24: **for each** $v \in v_s$ **do**
25:               **if** $v_s \neq \varphi$ **then**
26:                   select $w \in v_s \mid \overrightarrow{F}_{max} = \max_{w \in v_s} \overrightarrow{F}_{w, M_s} - \overrightarrow{E}_w$
27:                   Check for other destination in the direction of uw and verify whether they can also

---

be addressed by the same neighbour, w.

28:                    $SendQuery(w, M_s)$

29:              **end if**

30: **end for**

31: **end procedure**

5. SECTOR-BASED MULTICAST ROUTING ALGORITHM

In this section, we discuss the sector-based virtual-force-based multi-cast routing algorithm.

5.1. Motivation

Though MRAV uses virtual force technique to correctly identify the set of neighbours to whom the query message needs to be forwarded, the message complexity for making the computations is on the higher side. One of the reasons for higher overhead and energy consumption derives from the fact that the algorithm compares all the nodes in the neighbourhood for making the routing decision. However, most of the nodes queried are not in the general direction of communication. The number of nodes that need to be enquired about the effective force values can hence be reduced. At each node u, we need to identify the subset of N_u that are least likely to be part of the multicast route.

The primary challenge while trying to decide which nodes can be excluded for path computation comes from the uncertainty regarding the rest of the network.

5.2. Basic working

Unlike MRAV and [12], we divide the nodes around u into α sectors, with each sector covering an angle of $\theta = \dfrac{2 \cdot \pi}{\alpha}$. Here, α varies from 3 to 6 depending on the density of the region containing u. For $\alpha = 4$, we have $\theta = \dfrac{\pi}{2}$. For each of the α sectors containing some destination nodes, we select the appropriate neighbour in that sector to forward the packet. We compute the effective force on each of the k neighbouring nodes v_1, v_2, ..., v_k in the sector s, by taking into account the cumulative effect of dampening force($\overrightarrow{E_{v_i}}$), u and M_s, where $M_s \subseteq M$, M_s contains all the destination nodes in sector s. Once the appropriate forwarding node v_s for the sector s is determined, the packet is communicated to v_s along with M_s, which is the set of destination nodes to be handled in sector s.

When a source node s is interested in transmitting a packet to a multicast group M, it first sends a *Query* message to establish the multicast route. This *Query* message contains a pointer to indicate the multicast group and the set of nodes in M represented using bit array. The algorithm for sending the *Query* message is given in Algorithm 1.

When a node u decides on forwarding a packet to the destinations, it determines whether it has to split the multicast packet or not. The node determines whether all destinations are in the same general direction or not by computing the effective force from destinations, $\overrightarrow{F_{u,M}}$. If $\overrightarrow{F_{u,M}}$ and $\overrightarrow{F_{s,u}}$ are in opposing directions, then the node determines the appropriate neighbour in the forward direction to route the *Query* packet. The best way to determine whether all nodes are in the same general direction is to check whether the direction of force from each destination, $\overrightarrow{F_{d_i}}$, where $d_i \in M$, are within $\pm\frac{\theta}{2}$ angle from $\overrightarrow{F_{u,M}}$.

Figure 3 shows a simple illustration on how the force vectors are used in our algorithm. First, by using Equation 3, the effective repulsive force on the node u due to the destinations in multicast group M, given as $\overrightarrow{F_{u,M}}$, is computed. The next step is to divide the region around the node u into α sectors. Here α is the split parameter that we use. For dividing the region into sectors, we first look at the unit vector corresponding to the effective destination force, $\hat{F}_{u,M}$. From the direction of the unit vector, we take all nodes within angle $\pm\frac{\theta}{2}$ as the first sector, where $\theta = \frac{2 \cdot \pi}{\alpha}$. In the example in Figure 3, $\alpha = 4$ and $\theta = \frac{\pi}{2}$. The nodes within the next θ radians form part of the second sector. Thus, α sectors will be marked around the current node u. The four sectors in Figure 3 are divided by the dotted lines as illustrated.

Now, we divide the neighbours of u into α sets (V_1, V_2,..., V_α), such that all neighbouring nodes in sector s are put inside the set V_s. We apply the same procedure to divide the nodes in M into (M_1, M_2,..., M_α), where each set M_s indicates the set of destination nodes in the current destination set M. In the example in Figure 1, the set M is divided into two sets $M_1 = \{d_1\}$ and $M_2 = \{d_2, d_3, d_4\}$ corresponding to sectors 1 and 2. Similarly, the neighbours of u are divided into two sets; $V_1 = \{v_{n1}, v_{n2}\}$ and $V_2 = \{v_{n3}, v_{n4}, v_{n5}\}$. Here, $\bigcup_i V_i = N_u$, where N_u is the set of neighbouring nodes of the current node u. Also, $\bigcup_i M_i = M$.

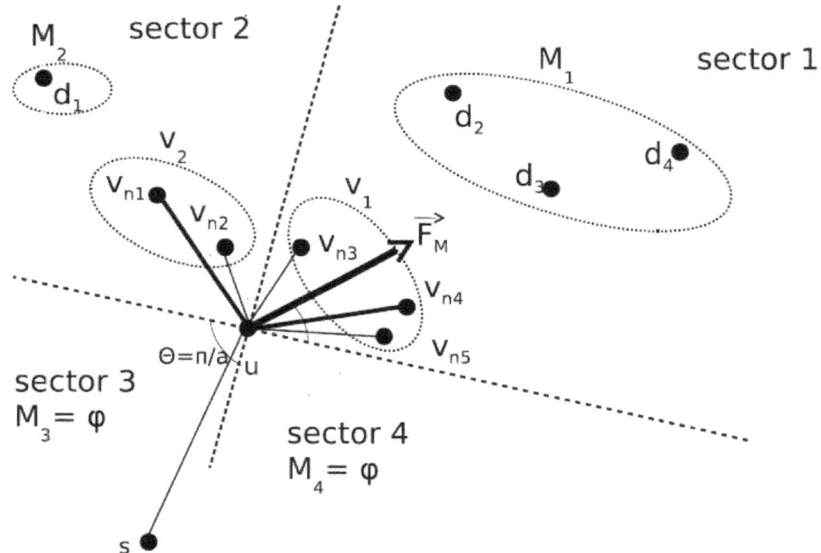

Figure 3. Virtual Force acting on node u with $\alpha = 4$.

Once the subsets of M, M_s, has been computed, we can easily determine whether any sector is having destination or not by checking for the condition: $M_s = \varphi$. If the condition holds true, then we don't have to explore that sector. If $M_s \neq \varphi$, then we have to determine appropriate neighbour to forward the packet in that sector s.

To choose the appropriate neighbour in sector s, first we have to determine whether $v_s = \varphi$. If so, then we have encountered a void in that sector. The simplest way to overcome void in the network is to follow the approach taken by Liu, *et al.* in [7]. Virtual force is used to cross across the void if possible, or else the perimeter rule of greedy perimeter stateless routing (GPSR) algorithm [13] is applied.

If $v_s \neq \varphi$, then the effective force on the packet as given in Equation 4 is computed for each neighbour in v_s. The neighbour $w \mid w \in v_s$ with the maximum force on the packet is chosen as the next hop neighbour.

5.3. Algorithm

The algorithm that uses the notion of sectors and virtual force, sector-based virtual force-based multicast routing algorithm (VFM), is shown in Algorithm 2.

Algorithm 2: Virtual Force-based Multicast routing algorithm

1: **procedure** SendQuery(u,M)
2: if $u \in M$ then
3:      $M \leftarrow M - \{u\}$
4: end if
5: for each $v \in N_u \wedge v \in M$ do
6:      Call $SendQuery(v, M)$
7: end for
8: $\overrightarrow{F_{u.m}} \leftarrow calculate VirtualForce(u, M)$
9: Divide region into α sectors centred at u, each of angle θ such that first sector lies within $\pm\frac{\theta}{2}$ of $\hat{F}_{u.M}$
10: for each sector s do
11:         $v_s \leftarrow \varphi$
12:         for each $v \in N_u$ do
13:             if $s = sector(v)$ then
14:                 $v_s \leftarrow v_s \cup v$
15:             end if
16:         end for
17: end for
18: for each sector s do
19:         $M_s \leftarrow \varphi$
20:         for each $d \in M$ do
21:             if $s = sector(d)$ then
22:                 $M_s \leftarrow M_s \cup d$
23:             end if
24:         end for
25: end for

26: for each sector s such that $M_s \neq \varphi$ do

27:          if $\gamma_s \neq \varphi$ then

28:              select $w \in \gamma_s \mid \overrightarrow{F_{max}} = \max_{w \in \gamma_s} \overrightarrow{F_{w,M_s}} - \overrightarrow{E_w}$

29:                  $SendQuery(w, M_s)$

30:          **end if**

31: **end for**

32: **end procedure**

6. SIMULATION RESULTS

We have performed simulation of our algorithm using a simulator written in Python language. The simulator uses the same energy model and Two Ray propagation model as implemented in ns-2. For the simulation runs, we have placed nodes in a fixed area of 2000m x 2000m with maximum transmission range set as 250m. We have compared MRAV and VFM with the QBIECRA algorithm proposed by Rahman, *et al.* [11]. While comparing with the sector-based VFM algorithm, we have taken α values as 3, 4 and 6. We have chosen QBIECRA since it uses a notion of quadrants for computing the multicast tree. Though Rahman, *et al.* also proposed another version of their quadrant based algorithm in [12], while performing simulations we didn't perceive a major difference between the results of QBIECRA and 4-N Intelligent routing. This is may be due to the fact that these algorithms only differ on the basis of which four neighbours are going to be chosen for computing the next forwarding node in the multicast path. The simulation results obtained are discussed in the rest of this section.

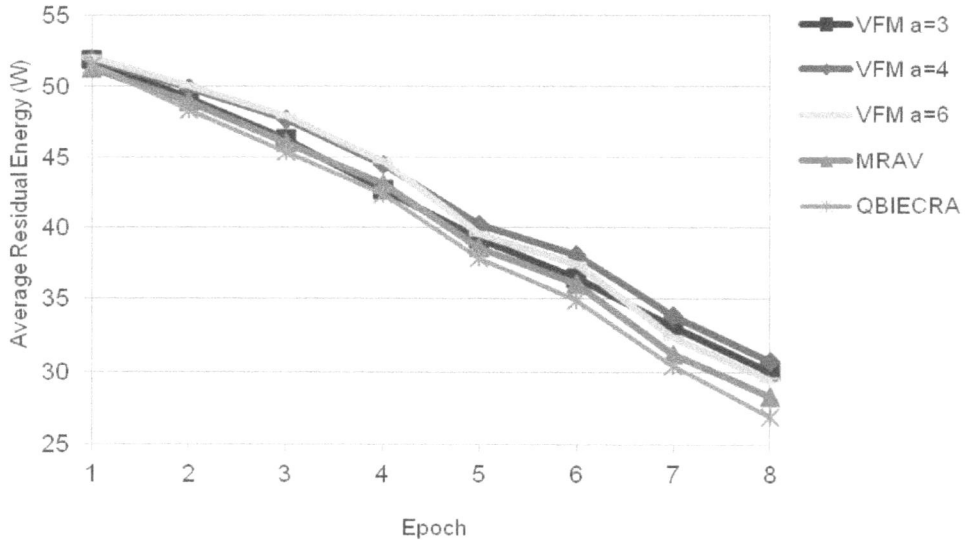

Figure 4. Average normalized residual energy over time

Average normalized residual energy is computed as the average residual energy of all nodes in the network normalized in percentage terms to account for variations in initial energy values of various nodes. We have computed the average of residual energy in each of the nodes after completing one cycle of packet transmissions for a sender node in each epoch, and are showing the normalized value in percentage terms after each epoch. The results for normalized average residual energy are shown in Figure 4. MRAV performs slightly better than QBIECRA. We observed that in the initial period, the choices made by QBIECRA and MRAV were same in many cases. But the choices began to diverge only later, when there was significant difference between the choices made, primarily due to the effect of the dampening force. For VFM, we observed that for all values of α, it performed as same as or better than QBIECRA and MRAV.

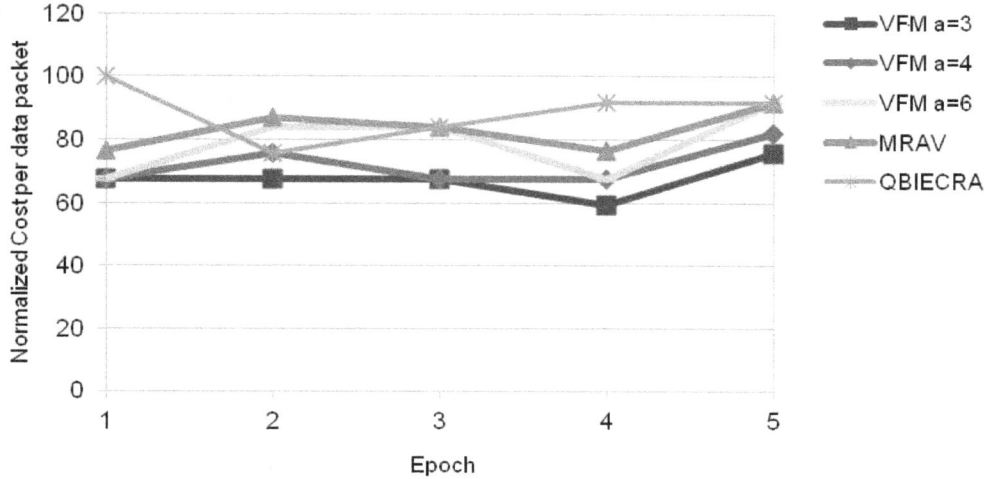

Figure 5. Normalized cost per data packet over time

Normalized cost per data packet is computed as the overall cost for sending a data packet from source node to destinations that is normalized in terms of percentile energy cost for the sake of comparison across multiple simulation runs with varying network sizes. The results for this metric are shown in Figure 5. We observed that VFM performs better than QBIECRA primarily by optimizing the effective energy spent on optimal path computation. Due to the number of comparisons made, MRAV was performing worse than VFM. We couldn't statistically establish a clear difference between MRAV and QBIECRA based on normalized cost per data packet.

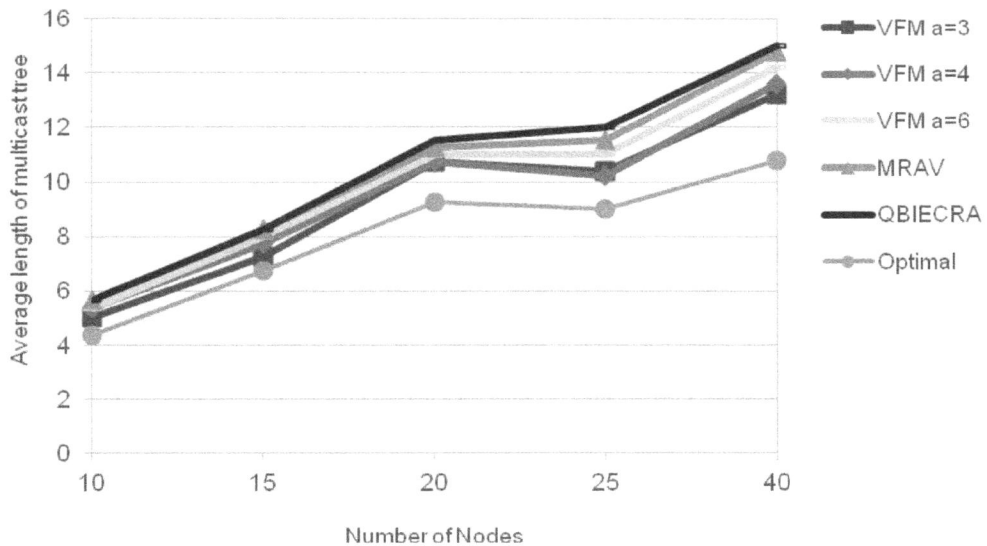

Figure 6. Average length of multicast tree with respect to number of nodes

We also observed the effective length of the multicast tree generated as shown in Figure 6. We observed a variation in the general trend of length of the multicast tree for 20 nodes for both of our algorithms. We attribute this variation to the choice of multicast set, presence of voids and other aspects specific to the test graphs used for simulation. Even with this variation, both MRAV and VFM provided shorter length for multicast tree as compared to QBIECRA. MRAV performed consistently better than QBIECRA as far as this metric is concerned for a larger network. For very small networks, there was no major difference between the two algorithms. We observed that 120° is the best angle between the branches for getting relatively minimal Steiner trees as defined by Gilbert, *et al.*[2]. As expected, the length of the multicast tree was nearing the optimal values for $\alpha=3$. We surprisingly found that we were getting good results for $\alpha=4$ as well. This was primarily due to the fact that we were not having many out-going edges in the opposite direction of the forward path. When it came to $\alpha=6$, our algorithm performed worse as expected, as we were generating too many out-going edges in some nodes resulting in a sub-optimal multicast tree. However, the performance of VFM algorithm was still better than QBIECRA as seen from the figure.

7. CONCLUSIONS

We have applied the notion of virtual force for energy efficient multicast routing in MANETs. We have presented two algorithms centred on the virtual-force technique. Our algorithm generates relatively minimal Steiner trees for use in multicast routing. The simulation results indicate that our algorithm performs better than other quadrant/sector based multicast routing algorithm. We have evaluated appropriate choice for the number of sectors to be used, and have found that $\alpha = 3$ or 4 can be used to generate energy efficient multicast paths. While we were applying the notion of virtual forces for multicasting, we observed that the approach as a whole provides good results. We haven't yet explored the relationship between the value of α and network density and spread. However, we believe that there is still scope in fine tuning the force model used. Though we had only used life of the network for QoS, the dampening force can easily be extended to include other parameters as well. The proposed algorithms have shown that the virtual force approach can be successfully used in MANETs. To our knowledge, this is the first paper that performs multicast routing algorithm using the virtual force approach in mobile ad-hoc networks. This is also the first paper to perform virtual-force computation on the basis of sectors.

REFERENCES

[1]   Luo Junhai, Ye Danxia, Xue Liu, and Fan Mingyu. A survey of multicast routing protocols for mobile ad-hoc networks. *Communications Surveys Tutorials, IEEE*, 11(1):78 –91, quarter 2009.

[2]   E. Gilbert and H. Pollak. Steiner minimal trees. *SIAM Journal on Applied Mathematics*, 16(1):1–29, 1968.

[3]   Marshall Bern and Paul Plassmann. The Steiner problem with edge lengths 1 and 2. *Information Processing Letters*, 32(4):171 – 176, 1989.

[4]   Glencora Borradaile, Philip Klein, and Claire Mathieu. An o(n log n) approximation scheme for steiner tree in planar graphs. *ACM Trans. Algorithms*, 5(3):31:1–31:31, July 2009.

[5]   S. Poduri and G.S. Sukhatme. Constrained coverage for mobile sensor networks. In *Robotics and Automation, 2004. Proceedings. ICRA '04. 2004 IEEE International Conference on*, volume 1, pages 165 – 171 Vol.1, april-1 may 2004.

[6]   Y. Zou and Krishnendu Chakrabarty. Sensor deployment and target localization based on virtual forces. In *INFOCOM 2003. Twenty-Second Annual Joint Conference of the IEEE Computer and Communications. IEEE Societies*, volume 2, pages 1293 – 1303 vol.2, march-3 april 2003.

[7]   Cong Liu and Jie Wu. Virtual-force-based geometric routing protocol in manets. *Parallel and Distributed Systems, IEEE Transactions on*, 20(4):433 –445, april 2009.

[8]   Cong Liu and Jie Wu. Swing: Small world iterative navigation greedy routing protocol in manets. In *Computer Communications and Networks, 2006. ICCCN 2006. Proceedings.15th International Conference on*, pages 339 –350, oct. 2006.

[9]   Martin Mauve, Holger Füssler, Jörg Widmer, and Thomas Lang. Position-based multicast routing for mobile ad-hoc networks. *SIGMOBILE Mob. Comput. Commun. Rev.*, 7(3):53–55, July 2003.

[10]   S. Fotopoulou-Prigipa and A.B. McDonald. Gcrp: geographic virtual circuit routing protocol for ad hoc networks. In *Mobile Ad-hoc and Sensor Systems, 2004 IEEE International Conference on*, pages 416 – 425, oct. 2004.

[11]   F.M. Rahman and M.A. Gregory. Quadrant based intelligent energy controlled multicast algorithm for mobile ad hoc networks. In *Advanced Communication Technology (ICACT), 2011 13th International Conference on*, pages 1298 –1303, feb. 2011.

[12]   F.M. Rahman and M.A. Gregory. 4-n intelligent manet routing algorithm. In *Australasian Telecommunication Networks and Applications Conference (ATNAC), 2011*, pages 1 –6, nov. 2011.

[13]   Brad Karp and H. T. Kung. Gpsr: greedy perimeter stateless routing for wireless networks. In *Proceedings of the 6th annual international conference on Mobile computing and networking*, MobiCom '00, pages 243–254, New York, NY, USA, 2000. ACM.

8

GRID BASED WIRELESS MOBILE SENSOR NETWORK DEPLOYMENT WITH OBSTACLE ADAPTABILITY

Mr. Mayur C. Akewar[1] and Dr. Nileshsingh V. Thakur[2]

[1]Department of Computer Science & Engineering, Shri Ramdeobaba College of Engineering and Management, Nagpur University, Nagpur, India
mayurakewar12@gmail.com
[2]Department of Computer Science & Engineering, Shri Ramdeobaba College of Engineering and Management, Nagpur University, Nagpur, India
thakurnisvis@rediffmail.com

ABSTRACT

Mobile Sensors find their target position and placed themselves over the target field to achieve a certain goal in self deployment with certain additional functionality like sensor relocation. An efficient deployment scheme guaranteed maximum coverage with full connectivity. Certain variance of coverage could be manageable but loss of connectivity because of failure of some sensor node causes complete isolation of some sensing area which causes loss of data of that area. So it is not enough to be just connected, it should be K connected (K>1) to control the isolation. Again the obstacles over the target field should be managed during deployment. A self deployment scheme using mobile sensors is proposed which achieve K connectivity. The target area is divided into n × n square grid. The distributed algorithm deploys a sensor in each square grid cell to maximize the coverage and achieve K connectivity. A square grid deployment pattern is used which guarantee at least (K=4) connectivity. The first algorithm assumed a plane target area with no obstacle. The second proposed algorithm has the capacity of obstacle adaptability which assumed a target area with some obstacle. Simulation result shows maximum coverage with minimum moving cost and obstacle adaptability.

KEYWORDS

Wireless Sensor Network, Wireless Network, Mobile Sensor Network, Coverage, Connectivity, Movement Cost, Obstacle Adaptability

1. INTRODUCTION

Wireless Sensor Network (WSN) has emerged as an efficient technology for wide variety of applications such as home automation, military application, environmental monitoring, habitat monitoring etc. It consists of distributed autonomous sensors to monitor physical or environmental conditions and to corporately pass their data through the network to main location. Sensor nodes are severely constrained in terms of storage resources, computational capabilities, communication bandwidth and power supply. WSN face many challenges due to these many constraints such as life of network, efficiency and the performance of network. An efficient deployment technique handles these challenges. Deployment is very critical issue because it decides the performance of wireless sensor network. Given set of sensors are deployed with a goal of maximizing the area that covered by the sensors. The deployment can be deterministic [1] or random [11] or incremental [2] or self deployment [4-6], [8], [12], [13], [15]. In this paper, following self deployment problem has been considered: *"Given a target sensing field with an arbitrary initial sensor distribution, how should these sensors self organize into a connected ad hoc network that has the maximum coverage, at the cost of a minimum moving distance?"*

Uniform distribution of sensor nodes over the target field is very important to cover maximum area. For uniform distribution the node has the additional capability of locomotion such that the sensor moves from high dense area to low dense area. After random deployment, the node self deploy themselves to achieve uniform distribution. Node find its own location and self deploy it to desired location with the capability of mobility. With the help of self deployment the sensor relocate them to the low dense area to achieve maximum coverage. Self deployment is very efficient form of deployment to manage the network dynamics. Self deployment uses mobile sensors for placement of sensor node. Mobility includes different functionality in wireless sensor network like coverage optimization, better lifetime of network, better use of resources and relocation [4]. Sensor nodes may change their location after initial deployment. Mobility may apply to all nodes or only to subsets of nodes. M. Marks [26] give taxonomy of mobility. Mobility may be active or passive. In active mobility the sensors are intelligent enough to find their path and move while in passive mobility the sensors may move by human or environmental assistance. J. Luo and et al [24] defined two types of mobility; they are U- mobility and C- mobility called as double mobility. C- mobility is controllable while U-mobility is uncontrollable. C- mobility is the property of sensor node and U- mobility is affected by environmental condition like wind. Mobility is essential for self deployment [4], [5], [6], [8], [12], [13], [15], that is to find the position and move to deploy them. Two mobile sensor platforms Racemote [9] and Robomote [22] are used for mobile sensor network deployment. A wireless mobile sensor network deployment approach is proposed by handling the issues of coverage, connectivity and movement cost of sensor node.

In this paper, distributed algorithm which solve the above problems has been proposed and get the K connected deployment scheme after random deployment. The proposed approach used square deployment pattern for at least K=2 coverage and at most K=4. The sensor nodes should self deployed them after initial distribution of sensor nodes on the target area. We consider a square target area and divided it into square grids. After random deployment the sensor should self deploy them such that each grid cell has exactly one node and each node should connect with at least 2 sensor nodes. We try to reduce the communication overhead to only communicate the sensor nodes within the grid cell and also to reduce the movements of sensor nodes to save the energy which results in increasing network lifetime. The first algorithm assumed a plane target area with no obstacle. In most of the existing work researcher assumed a target field with no obstacle. The second proposed algorithm has the capacity of obstacle adaptability which assumed a target area with some obstacle. The proposed algorithm achieved a self deployment of sensors with minimum moving cost of sensors for minimizing energy consumption with obstacle adaptability. Simulation result shows maximum coverage with minimum moving cost and obstacle adaptability.

This paper is organized as follows: Section 2 gives the related work. Section 3 gives the preliminaries of the system. Section 4 gives the proposed scheme. Section 5 gives experimental results. The paper ends with conclusion and future work which is given in section 6.

2. RELATED WORK

Different approaches have been proposed for the deployment of mobile sensor for solving above problem. Some proposed approaches used virtual force [27] and [29] approach which considered the mobile nodes as a particle under attractive or repulsive force. The sensor nodes modelled as points subject to attractive or repulsive force according to the distance between two sensors. The virtual force exerted on the sensor is similar to the Coulomb force. By setting the threshold distance between sensors, each sensor moves in accordance with the summation of the force vectors and eventually a uniform deployment is achieved. The potential force [4] based method which used the method to migrate the node from high density to low density area.

Computational Geometry based approach which used the popular data structure Voronoi diagram has been proposed in [27]. According to F. Aurenhammer [30], it is believed that the Voronoi diagram is a fundamental construct defined by a discrete set of points. Voronoi diagram is used to discover the e Voronoi diagram is constructed on the basis of neighbors who are in the communication range. If there is less number of nodes in the communication range it will result in improper voronoi diagram construction. Second problem with this approach is that we will not get optimal deployment. Again existing approaches have the problem of optimal deployment and high communication overhead to communicate to construct the voronoi diagram or to calculate the virtual forces. There is chance of unnecessary movements of nodes in the existing approaches. Most of the existing studies mainly considered the issue of coverage, but sensor network is often gets into sensor failure so the sensor should be at least K connected with K>1. Existence of coverage hole once all the sensors have been initially randomly deployed in the target area. A node needs to know the location of the neighbors to construct the voronoi diagrams. The diagram partition the whole target area into voronoi polygons. Each polygon has a single node with the property that every point in this polygon is closer to this node to estimate any local coverage hole. After each iteration, every sensor node moves to an improved location and then the voronoi diagrams are reconstructed. Another structure that is directly related to Voronoi diagrams is the Delaunay triangulation [29], [13]. The Delaunay triangulation can be obtained by connecting the sites in the Voronoi diagram whose polygons share a common edge. It has been shown that among all possible triangulations, the Delaunay triangulation maximizes the smallest angle in each triangle. N. Azlina and *et al* [8] proposed coverage optimization algorithm based on particle swarm optimization (PSO). C. Hsien and *et al* [13] and S. Megerian and *et al* [14] proposed Delaunay triangle based approach. The algorithm eliminates the coverage holes near the boundary of sensing area and obstacles Delaunay triangle is applied for the uncovered regions. G. Tan and *et al* [6] proposed enhanced form of Virtual force method called as connectivity preserved virtual force method (CPVF). The authors consider the obstacles during deployment. The given approach is used for mobile sensor network for the self deployment. The authors overcome the limitations of previous approaches where large communication range, dense network and obstacle free field or full knowledge of the field layout have been assumed. M. Garetto [15] proposed a virtual force method used for sensor relocation. Upon occurrence of physical phenomenon nodes relocate themselves so as to control the event, while maintaining network connectivity. After event ends all nodes return to the monitoring configuration. J. Lee and *et al* [4] proposed a potential field based approach. After initial deployment, group of mobile sensor clusters are formed using potential field method and then cluster heads are used to establish hexagonal structure or achieving coverage. L. Filipe and *et al* [2] proposed an efficient incremental deployment algorithm using Largest Empty Circle problem. Algorithm indicates position for new nodes to be deployed and number of new nodes. A network is said to have k-coverage if every point in it is covered by at least k sensors. the k-coverage map for a square grid and THT. S. Shakkottai and *et al* [18] proposed unreliable sensor grids by considering issue of coverage and connectivity. X. Bai and *et al* [20] investigate different deployment models with their performance evaluation. B. Liu and *et al* [21] discuss he question "Is mobility improves coverage of sensor network?". S. Yang and *et al* [23] proposed scan based movement assisted deployment method. By handling the issue of coverage an algorithm named SMART (Scan based movement assisted) is proposed. A unique problem called communication hole is handled here. The algorithm control the moving distance. G. Wang and G. Cao [27] proposed Coverage hole detection method. A robot deployment algorithm that overcomes obstacle and employs full coverage with minimal number of sensor node is discussed in [10]. A mobile Robot is used for placing sensors. The robot explores the environment and deploys a stationary sensor to the target location from time to time.
Non uniform sensor deployment in mobile sensor network is discussed in [16]. A Jagga and *et. al.* [31] proposed an approach for deploying mobile agents. G. Wang and *et. al.* [32] proposed an approach for dynamic deployment by using both static and dynamic sensor nodes.

Most of the work evaluates the performance of the network on the basis of coverage and number of sensors. Use of mobility for sensor relocation to improve coverage, and network lifetime after failure of some nodes in the system has proposed in some work. Constrained coverage approach for better network performance and energy efficiency has been proposed in some work. Balancing energy consumption by placing sensors in terms of different densities (non uniform deployment) is proposed. Computational geometry data structure like voronoi diagram and Delaunay tringulation is used for finding coverage hole is proposed in most of the work. Virtual force method and movement assisted method is used for movement of sensors in most of the work. Coverage is an open issue which can be handled in most of the approach. Sensor movement also need energy, so cost of sensor movement should be minimized. Efficient management of mobility towards a better coverage remains an unanswered question.

We are using a square deployment pattern and a grid based approach such that we can manage the mobility of the sensor node efficiently. The cost of the movement is minimized considerably in our proposed approach. We achieved nearly hundred percent coverage with maximum K=4 connectivity. So our approach is different from existing approach such that we achieve both maximum coverage and K connectivity with minimum movement cost. Also the obstacle over the target field is handled in a better way.

3. PRELIMINARIES

3.1. System Assumptions

We assume that the target area is on a 2D plane. We assume that all the sensor nodes are homogenous with same sensing and communication range. Both ranges are assumed as disk based model. If the two sensors are in range then they are called as neighbors. A sensor can determine the position of other sensor only by communication. The sensing range R_s is equal to $\sqrt{2} \times$ the length of the side of square grid to achieve optimal coverage and communication range R_c is equal to $\sqrt{2} \times R_s$ to have a connected network. Each sensor node knows the coordinates of their position by using some localization technique like GPS. Sensor nodes should be able to change their position coordinates after each movement. We may use a mobile sensor node platform Robomote [28] and Racemote [9]. Each sensor node knows the dimension of square target area, also each sensor node knows the dimension of square grid which is fixed for each cell and should be able to get the centre coordinates of the square cell where it is currently deployed. Sensor node moves with uniform speed horizontally or vertically in a straight line for fixed amount of time which we called as time.

3.2. Target Field

We are using a square target field on 2D plane surface. The target field is divided into equal size square grid cell. The field is divided into 3×3 grid. Each cell has given a cell id, starting from left upper corner. Cell id's are from 1 to 9.

3.3. Deployment Pattern for Achieving K Connectivity

We have divided the target area into equal length square grids cell to achieve optimal coverage and K connectivity. Reason of using an optimal deployment pattern [1] is that the sensor network could subject to failures, so the sensor node should at least K connected (K>1). If anyone node failed, then other node could serve the purpose. A square grid provides at least 4- connectivity [6]. When $1.14 \leq R_c/R_s \leq \sqrt{2}$, the square pattern is better than other deployment pattern [6]. We are assuming the sensing and communication range within the above range.

3.4. Information to Each Sensor

Each sensor knows the dimension of target field and the grid cell. Position of the sensor node is known with the help of GPS device. Also the node should get the centre of cell by using its position coordinates. Each cell only communicates to other node within its own cell. Each sensor node has a list of the connected nodes within the cell. By which the nodes within the cell decides the next action by local communication. Each senor node has a list of closest boundary distance and its direction to decide the next move. Each node has an energy counter which is decremented after each movement to calculate the movement cost of the sensor. Each sensor node should move until the goal is reached or to a fixed amount of distance and decides the next action. Each sensor should identify the boundary of the obstacle. After identification of obstacle boundary the sensor node should move towards next closest boundary. Each node has a counter for time which set to some threshold value. After random deployment the node starts the timer and wait up to set threshold value. If it migrate to another cell then the timer again set to zero and again start the timer until another event is occur or to a threshold time. Each cell node has a cell id for identification of the cell in which it is currently deployed. The cell cannot return to the previous cell by searching the cell id list.

3.5. Movement of Sensor Node

Each sensor node moves for a fixed amount of time and distance in horizontal or vertical direction. After each move the energy is lost which is represented by an energy counter set to some value. For minimizing the movement cost the sensor node with maximum energy will move and the sensor node with minimum energy stays in the cell. Also the sensor will move towards closest boundary to minimize the movement cost. The sensor node cannot move to the same cell again by looking at the cell id list. For this, each sensor node has the history of cell id.

4. PROPOSED APPROACH

4.1. Problem Definition

Given a target sensing field with an arbitrary initial sensor distribution, these sensors should self organize into a k-connected ad hoc network that has the maximum coverage, at the cost of a minimum moving distance.

In the initial phase, the target area is divided into n × n equal square grids and the R_s, R_c values are decided on the basis of length of the side of the grid cell. $R_s = \sqrt{2} \times$ side of the cell and $R_c = \sqrt{2} \times R_s$. The sensor nodes are fixed which is equal to number of cells. If there are 9 cells then the nodes are also 9. The sensor nodes are randomly deployed on the target field. A distributed algorithm is executed in each sensor node to self deploy the sensor node such that each cell reach the center of the grid to achieve optimal deployment. The proposed approach gives maximum coverage with minimum moving distance and obstacle adaptability.

4.2. Scenario 1

If after random deployment, each cell has exactly one sensor node then the self deployment is quite simple. The list of connected nodes is empty. So the sensor node decides that my cell does not contain any other node, so the there has not any redundant node for migration or relocation. The sensor node knows the center of the current deployed cell, so the particular sensor just has to move to the center of the cell. Each sensor node will performed following steps.

1. All the nodes broadcast hello message with their position coordinates and wait for acknowledgement and start the timer.
2. If the node j present in the same cell then from position coordinates it will decides that the node i is in its own cell, it send acknowledgement message and they will connect if distance $(i, j) \leq Rc$.
3. If the node does not receive any acknowledgement means the cell contains only one node
4. After timer reaches a threshold value the node stop waiting and it will then execute the MovetoCenter Algorithm to reach to center of cell.

MovetoCenter algorithm is given below. Consider that a sensor node has the coordinate value Xnode and Ynode. The sensor will move until it reach the center of the cell to maximize the coverage, that is (Xnode, Ynode)=(Xcenter, Ycenter), where Xcenter and Ycenter are the x coordinate and y coordinate of center of a cell.

Algorithm
1. do{
2. If (Xnode< =Xcentre) && (Ynode<=Ycentre)
3. {Xnode=Xnode+(Xcentre, Xnode);Moveto(Xnode, Ynode);
4. Ynode= Ynode+(Ycentre- Ynode);Moveto(Xnode, Ynode);
5. Else If (Xnode< =Xcentre) && (Ynode>=Ycentre)
6. {Xnode=Xnode+(Xcentre-Xnode); Moveto(Xnode, Ynode);
7. Ynode= Ynode-(Ynode- Ycentre); Moveto(Xnode, Ynode);}
8. Else If (Xnode>=Xcentre) && (Ynode>=Ycentre)
9. {Xnode=Xnode-(Xnode-Xcentre); Moveto(Xnode, Ynode);
10. Ynode= Ynode-(Ynode- Ycentre); Moveto(Xnode, Ynode);}
11. Else If (Xnode> =Xcentre) && (Ynode=<Ycentre)
12. { Xnode=Xnode-(Xnode-Xcentre); Moveto(Xnode, Ynode);
13. Ynode= Ynode+(Ycentre- Ynode); Moveto(Xnode, Ynode);}
14. }while((Xnode, Ynode)==(Xcenter, Ycenter));

4.2.1. Execution of Scenario 1

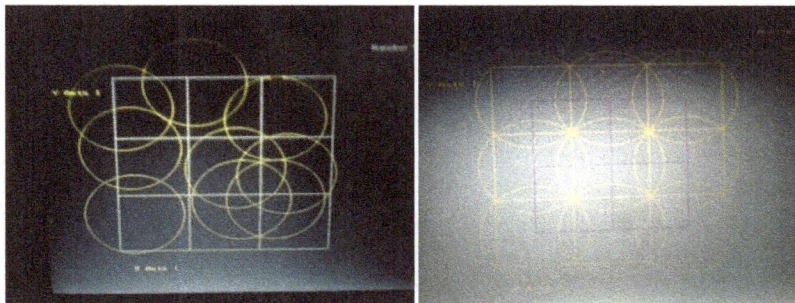

Figure 1 Initial and Final Deployment

4.3. Scenario 2

After random deployment if there is more than one sensor is deployed in a cell then it will follow following strategy.

a) Identifying Redundant Node: If a cell contains more than one node after initial distribution then each node has to decide the nodes which are redundant. The redundant nodes migrated to other cell which is close to it. Foe identification of redundant nodes, we use some distance measures. For each cell we have decide a fixed location. The distance between that location and the location of node is calculated. The node with minimum distance stayed in the cell and the other node will move towards the closest boundary until they will migrate to new cell. If any cell has only one node then that node will move to a threshold distance such that if another node comes then it should be connected to that node.

b) Movement of Node: The redundant nodes will move towards closest boundary until migrated to new cell and updated their position coordinates. The direction for movement is towards closest boundary. If the closest cell is the cell from where the node was coming, in such a case the node will move towards the next closest boundary. For this purpose the node look at the history of cell id.

c) Minimize movement cost: Each sensor node has two counter one is timer 't' and other is migration counter 'm', each one is set to zero. If there is only one node within a given cell then the counter t is started until a threshold time. After that time is reached then the sensor decides that know sensor is migrated in its cell then the node will follow my MovetoCenter algorithm to move to the center. If a node is migrated then the other node and the migrated node set t=0 and decides the next move by communication. The migrated node increment the counter to 1 and push the other node to migrate to other cell which is depends upon the distance between the boundary and the location of the sensor. If there is a tie then the sensor node with maximum distance from center to its location moves and other stays. After each communication within the cell the node set the timer to t=0. The nodes which are deployed on the boundary cell are quite intelligent to decide not to move towards the boundary of the target field.

4.3.1. Steps for Scenario 2

1. All the sensor nodes broadcast hello message with their position coordinates and wait for acknowledgement and start the timer.
2. If the nodes gets acknowledgement then the cell contains more than one node and they update their connected node list.
3. If the nodes does not get any acknowledgement means the node is isolated, then the nodes will move towards center until it is connected or reach a threshold distance.
4. The nodes calculate the distance from a fixed location which is depends upon the cell.
5. The nodes share their distance list through local communication to all the nodes in the given cell.
6. The node compares the distance.
7. The node with maximum or minimum distance (which is depends upon the cell) stay in the cell and the other (redundant sensors nodes) migrated to nearest cell and increment the migration counter, and update cell id and position coordinates. If the nearest cell has already visited then the sensor node will move towards the next nearest cell.
7. After migration the timer again set to zero and the node waits for local communication by sending hello message in some time interval.
8. If the migrated sensor node receives acknowledgement then it will repeat step 4 to 7.
9. If two sensor nodes has same migration counter then the nods which will migrated is decides on the basis of energy level. The node with high energy will migrate.
10. If the node has to move to the same cell again, then from cell id it will decides not to move to that cell. It then migrated to the next closest boundary.
11. If each cell contains exactly one node and timer=threshold time then the MovetoCenter algorithm will execute.

4.3.2. Execution of Scenario 2

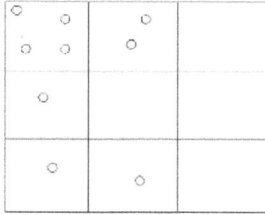

Figure 2 Initial Phase

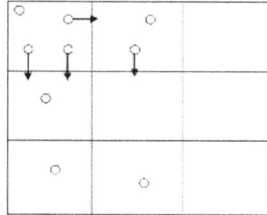

Figure 3 Iteration 1

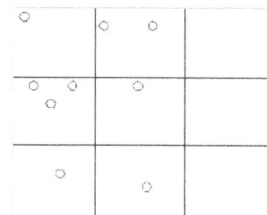

Figure 4 Iteration 2

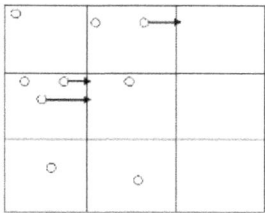

Figure 5 Iteration 3

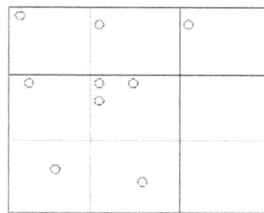

Figure 6 Iteration 4

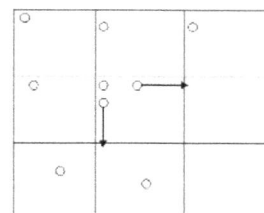

Figure 7 Iteration 5

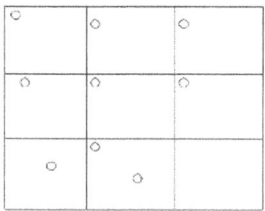

Figure 8 Iteration 6

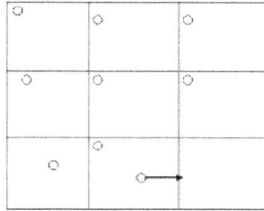

Figure 9 Iteration 7

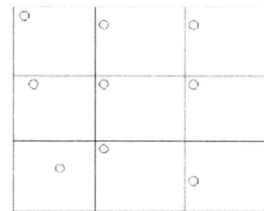

Figure 10 Iteration 8

4.4. Scenario 3

The target field contains any obstacle. Our previous algorithm will oscillate because at any given time the condition of all the cell contains exactly one node cannot satisfy.

4.4.1. Obstacle Adaptability

Sensor node knows the boundary of obstacle. During movement of sensor node if any obstacle arrives in the path of sensor node, then the sensor node will turn and move towards the next closest boundary. For achieving stability we set the migration counter to a threshold value and execute the previous algorithm. The sensor migrated only for the fixed count. If the migration count is equal to one then the node will migrate only once.

We are assuming that a cell is totally covered by an obstacle. So during movement of sensor, if the closest boundary is towards obstacle covered cell, then the sensor node will unable to move or migrate to that cell, so the node calculates the next closest boundary and move towards that cell.

We are considering that the obstacle is in center cell as shown in figure. After execution of the algorithm any cell contains more than one node. In such a case the sensor node with maximum energy will executes the MovetoCenter algorithm while the other node will go to the sleep mode, which could be used for the incremental deployment if the node in the cell has failed.

Figure shows the movement of sensor if any obstacle comes in the path of sensor node.

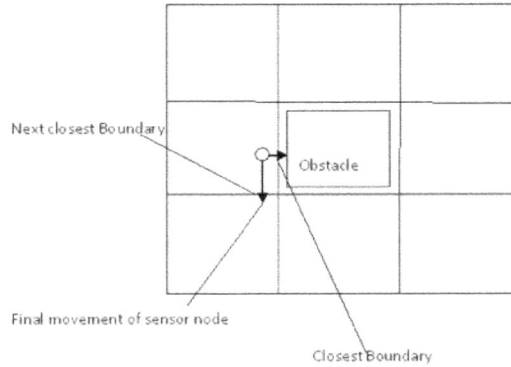

Figure 11 Obstacle Adaptability

4.4.2. Execution of Scenario 3

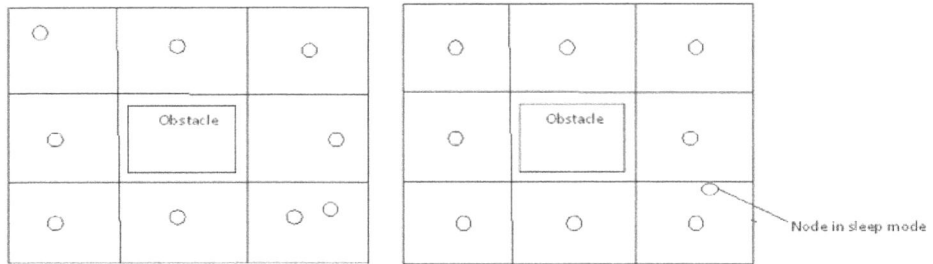

Figure 12 Initial and Final Configuration

5. EXPERIMENTAL RESULTS

We have used a 'C' based simulator for simulating the above mobile deployment scheme. We have considered a 3×3 grid square target field (300 × 300 meter square) with 9 sensors such as each cell must have 1 node deployed. Each cell has a area of 100 ×100 meter square. Sensing range is decided as Rs=71 meter and communication range as Rc=101 meter. Energy counter is initialized as 2000 and migration counter is initialized as m=0. The target area is fully covered with at most K=4 connected sensor node. The performance graph shows the energy lost during movement for various scenarios.

5.1. Execution 1

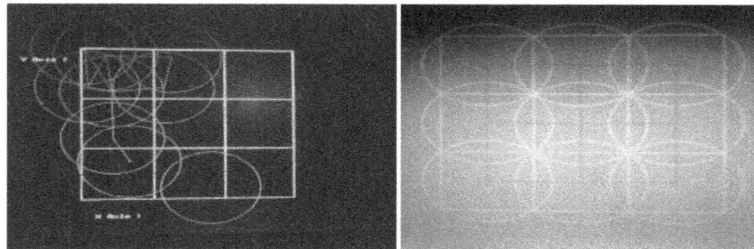

Figure 13 Initial and Final Configuration

Initially the field does not contain any obstacle. The surface is flat. Figure shows the initial configuration after random deployment. The distributed algorithm is executed and the final configuration is achieved with 100% coverage and at most K=4 connectivity. The performance graph shows the movement cost during deployment of the sensors. The circle shows the sensing range and the connectivity is shown by lines.

5.1.1. Performance Graph of Eecution 1

Graph shows the enegy lost during sensor movement.

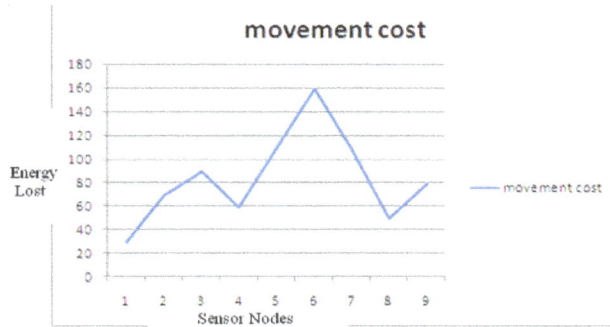

Figure 14 Cost of Node Movement

5.2. Execution 2

The target field contains obstacle. Figure shows the initial configuration after random deployment. The performance graph shows the movement cost during deployment of the sensors. The circle shows the sensing range. The middle cell contains obstacle.

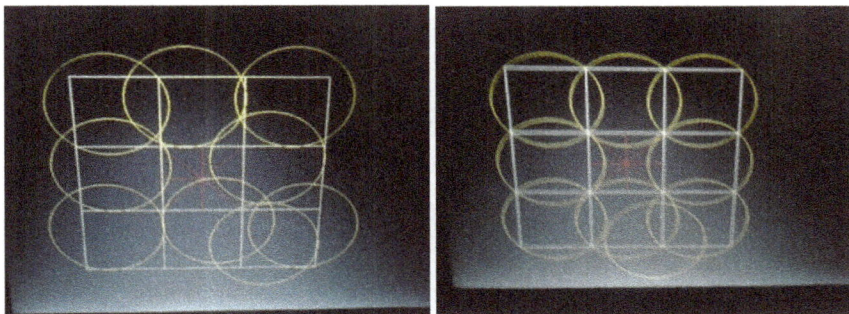

Figure 15 Initial and Final Configuration with Obstacle

5.2.1. Performance Graph of Execution 2

The graph shows the energy lost during movement of sensors. Performance is evaluated by considering different migration count. We are considering maximum migration count m=4.

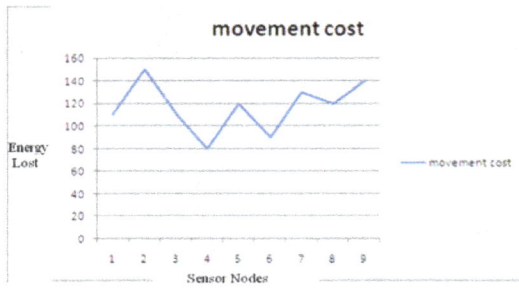

Figure 16 migration count=1

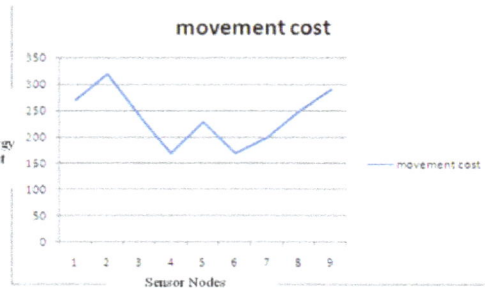

Figure 17 migration count=2

Figure 18 migration count=3

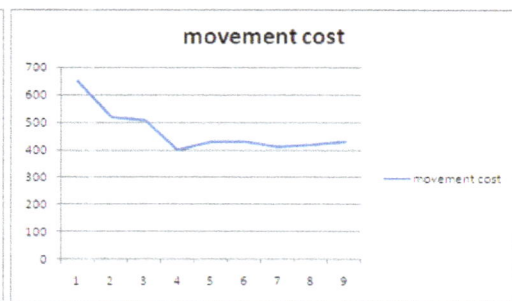

Figure 19 Migration count =4

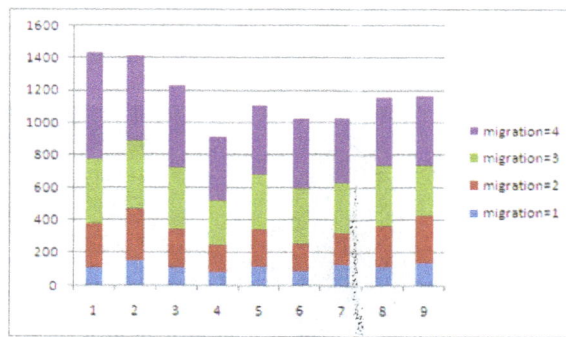

Figure 20 Energy lost for various migration counts

6. CONCLUSION AND FUTURE WORK

We proposed a grid based mobile deployment algorithm which self deploy the fixed number of nodes in each grid cell. The distributed algorithm is efficient to minimize the cost of movement and communication overhead. The square deployment pattern has at least 4-connectivity and coverage. In future work we can have more experiments on the obstacle avoidance strategy while deployment by considering obstacles at different position on the target area and measure the energy lost by setting the migration counter. We can use this 3×3 grid as a mask for a larger target area and large amount of sensor nodes. Also different deployment patterns like hexagonal and triangular pattern can be used for self deployment scheme.

REFERENCES

[1] Ziqiu Yun, Xiaole Bai, Dong Xuan, Ten H. Lai, and Weijia Jia "Optimal Deployment Patterns for Full Coverage and k-Connectivity (k<=6) Wireless Sensor Networks" IEEE/ACM transactions on networking, vol. 18, no. 3, June 2010.

[2] Luiz Filipe M. Vieira, Marcos Augusto M. Vieira, Linnyer Beatrys Ruiz, Antonio A.F. Loureiro, Di´ogenes Cec´_lio Silva, Ant.onio Ot´avio Fernandes "Efficient Incremental Sensor Network Deployment Algorithm".

[3] Kay Romer and Friedemann Mattern "The design Space of Wireless Sensor Networks" IEEE Wireless Communications, Dec. 2004.

[4] Jaeyong Lee, Avinash Dharne and Suhada Jayasuria "Potential Field Based Hierarchical Structure for Mobile Sensor Network Deployment" Proceedings of the 2007 American Control Conference, July 2007.

[5] W. Li, Y. I. Kamil and A. Manikas "Wireless Array Based Sensor Relocation in Mobile Sensor Networks" ACM 2009.

[6] Guang Tan, Stephen A. Jarvis, Member, and Anne-Marie Kermarrec "Connectivity-Guaranteed and Obstacle-Adaptive Deployment Schemes for Mobile Sensor Networks" IEEE transactions on mobile computing, vol. 8, no. 6, June 2009.

[7] Wint Yi Poe and Jens Schmitt "Node Deployment in Large Wireless Sensor Networks: Coverage, Energy Consumption, and Worst-Case Delay" ACM November 2009.

[8] Nor Azlina Bt. Ab Aziz , Ammar W. Mohemmed and Mohammad Yusoff Alias "A Wireless Sensor Network Coverage Optimization Algorithm Based on Particle Swarm Optimization and Voronoi Diagram" Proceedings of the 2009 IEEE International Conference on Networking, Sensing and Control, Okayama, Japan, March 26-29, 2009.

[9] G. Song, W. Zhuang, Z. Wei, and A. Song, "Racemote: A Mobile Node for Wireless Sensor Networks," in Proc. 5th IEEE Conference on Sensors, IEEE, 2006.

[10] Chih-Yung Chang, Jang-Ping Sheu, Yu-Chieh Chen, and Sheng-Wen Chang " An Obstacle-Free and Power-Efficient Deployment Algorithm for Wireless Sensor Networks" IEEE transactions on systems, man, and cybernetics—part a: systems and humans, vol. 39, no. 4, july 2009.

[11] Jiming Chen, Entong Shen and Youxian Sun "The deployment algorithm in Wireless Sensor Network: A Survey" Information Technology Journal 2009.

[12] Jun Li, Baihai Zhang and Lingguo Cui "A Restricted Delaunay Tringulation Graph Based Algorithm for Self-Deployment in Mobile Sensor Networks" Journal of Computational Information Systems 2010.

[13] Chun-Hsien Wu, Kuo-Chuan Lee, and Yeh-Ching Chung "A Delaunay Triangulation Based Method for Wireless Sensor Network Deployment".

[14] Seapahn Megerian, Farinaz Koushanfar, Miodrag Potkonjak, and Mani B. Srivastava "Worst and Best-Case Coverage in Sensor Networks", IEEE transactions on mobile computing, vol. 3, no. 4, october-december 2004.

[15] Garetto M, Gribaudo M, Chiasserini CF, Leonardi E (2007) "A distributed sensor relocation scheme for environmental control", In: The ACM/IEEE Proc. of MASS

[16] Cardei M, Yang Y, Wu J (2008) Non-uniform sensor deployment in mobile wireless sensor networks. In: Proc. Of WoWMoM, pp 1–8

[17] W. Li and C. G. Cassandras, "A minimum-power wireless sensor network self-deployment scheme," in *Proc. Annu. IEEE WCNC*, New Orleans, LA, Mar. 2005, vol. 3, pp. 1897–1902.

[18] Shakkottai, S., Srikant, R., and Shroff, N. Unreliable Sensor "Grids: Coverage, Connectivity and Diameter". In IEEE Twenty-Second Annual Joint Conference of the IEEE Computer and Communications Societies(INFOCOM 2003) (New York, NY, USA, Mar. 2003), vol. 2, ACM,pp. 1073–1083.

[19] Heo N, Varshney P (2005) "Energy-efficient deployment of intelligent mobile sensor networks", IEEE Trans Syst Man Cybern 35:78–92.

[20] X. Bai, S. Kumar, D. Xuan, Z. Yun, and T. H. Lai, "Deploying wireless sensors to achieve both coverage and connectivity," in *Proc. ACM MobiHoc*, 2006, pp. 131–142.

[21] B. Liu, P. Brass, and O. Dousse, "Mobility Improves Coverage of Sensor Networks," Proc. ACM MobiHoc, 2005.

[22] Dantu, K., Rahimi, M., Shah, H., Babel, S., Dhariwal, A., and Sukhatme, G. "Robomote: Enabling Mobility In Sensor Networks", IEEE/ACM International Conference Information Processing in Sensor Networks, Apr. 2005.

[23] S. Yang, M. Li, and J. Wu, "Scan-based movement-assisted sensor deployment methods in wireless sensor networks," *IEEE Trans. On Parallel and Distributed Systems*, vol. 18, no. 8, pp. 1108–1121, 2007.

[24] Ji Luo, DanWang, Qian Zhang, "Poster Abstract: Double Mobility: Coverage of the Sea Surface with Mobile Sensor Networks", *Mobile Computing and Communications Review, Volume 13, Number 1*

[25] Lingxuan Hu, David Evans "Localization for Mobile Sensor Networks", *MobiCom'04*, Sept. 26.–Oct. 1, 2004, Philadelphia, Pennsylvania, USA. 2004 ACM 1-58113-868-7/04/0009

[26] Michał Marks "A Survey of Multi-Objective Deployment in Wireless Sensor Networks", Journal of Information, 2010

[27] Wang G, Cao G, Porta TL (2006) Movement-assisted sensor deployment. IEEE Trans Mob Comput 6:640–652

[28] A Sekhar, B. S. Manoj, and C. S. R. Murthy, "Dynamic coverage maintenance algorithms for sensor networks with limited mobility," in *Proc. 3rd IEEE Int. Conf. Pervasive Comput. Commun. (PerCom)*, Kauai Island, HI, Mar. 2005, pp. 51–60.

[29] Poduri S, Sukhatme GS (2004) Constrained coverage for mobile sensor networks. In: Proc. of IEEE ICRA

[30] F. Aurenhammer, "Voronoi Diagrams—A Survey of a Fundamental Geometric Data Structure," ACM Computing Surveys 23, pp. 345-405, 1991.

[31] A. Jagga, Kuldeep, and V. Rana "A Hybrid Approach for Deploying Mobile Agents in Wireless Sensor Network", IJCA, 2012.

[32] G. Wang, L. Guo, H. Duan, L. Liu and H. Wang, "Dynamic Deployment of Wireless Sensor Networks by Biogeography Based Optimization Algorithm", *J. Sens. Actuator Netw.* 2012, *1*, 86-96; doi:10.3390/jsan1020086.

9

A BI-SCHEDULER ALGORITHM FOR FRAME AGGREGATION IN IEEE 802.11N

Vanaja Ramaswamy[1], Abinaya Sivarasu[2], Bharghavi Sridharan[3] and Hamsalekha Venkatesh[4]

[1, 2, 3, 4] Department Of Computer Science and Engineering,
Sri Venkateswara College Of Engineering,
Sriperumpudur taluk, Tamilnadu, India

ABSTRACT

IEEE 802.11n mainly aims to provide high throughput, reliability and good security over its other previous standards. The performance of 802.11n is very effective on the saturated traffic through the use of frame aggregation. But this frame aggregation will not effectively function in all scenarios. The main objective of this paper is to improve the throughput of the wireless LAN through effective frame aggregation using scheduler mechanism. The Bi-Scheduler algorithm proposed in this article aims to segregate frames based on their access categories. The outer scheduler separates delay sensitive applications from the incoming burst of multi-part data and also decides whether to apply Aggregated - MAC Service Data Unit (A-MSDU) aggregation technique or to send the data without any aggregation. The inner scheduler schedules the remaining (delay-insensitive, background and best-effort) packets using Aggregated-MAC Protocol Data unit (A-MPDU) aggregation technique.

KEYWORDS

IEEE 802.11n, Frame Aggregation, Scheduler, A-MPDU, A-MSDU.

1. INTRODUCTION

IEEE 802.11n is the most popularly used wireless *Local Area Network (LAN)* technology today. It is one of the superior standards when compared to its previous standards like IEEE 802.11 a/b/g through its ability to provide maximum of 600 Mbps [4] data rate. This is due to the enhancements provided in IEEE 802.11n like Multiple Input Multiple Output (MIMO) and channel bonding [1] in PHY (Physical) layer and other enhancements in Medium Access Control (MAC) layer. The MIMO technology in PHY layer uses multiple antenna elements to send multiple copies of the data there by significantly increasing the data rate. It uses 40 MHz channel for transmission. The MAC layer enhancements are frame aggregation, block acknowledgement, reduced inter frame spacing and allowing transmission in reverse direction. Recently, many 802.11n based products are popularly used in many user devices. But the performance of these products is not up to the expectation due to various constraints in the real time. The fairness of the IEEE 802.11n is not very effective in case of unsaturated traffic.

This paper proposes a scheduler for the MAC layer which divides the packet based on Access Category (AC) to provide different QoS services for different category of data. It also decides between whether to use frame aggregation or not based on the urgency of the data. This provides fairness to the traffic for off time traffic.

2.   RELATED WORK AND LITERATURE SURVEY

2.1   PHY and MAC layer enhancements

The important enhancements in IEEE 802.11n which improved the performance of the WLAN are briefed here.

2.1.1   Multiple Input Multiple Output (MIMO)

IEEE 802.11n uses MIMO technology to improve the data rate up to 600 Mbps [10]. MIMO uses multiple antennas for transmission and reception for transmission of the same data at phases. It defines different M X N configurations from 1 X 1 to 4 X 4 [5]. As the number of antennas for transmission and reception increases, the performance of the channel increases significantly. MIMO operates in two different modes: SDM and STBC.

- In Spatial Division Multiplexing (SDM), the output stream is divided into multiple sub streams which are then spatially multiplexed and transmitted from multiple antennas. At the receiving side, these streams are demultiplexed and recovered.
- In Space Time Block Coding (STBC), the same output stream is transmitted over multiple antennas simultaneously. At the receiving side, the multiple streams with different strength are compared to recover the original stream.

802.11	Freq (Hz)	Through put (Mb/s)	Max Bit rate (Mb/s)	Modulation	r_{in} (m)
A	5	23	54	OFDM	~35
B	2.4	4.3	11	DSSS	~38
G	2.4	19	54	OFDM	~38
N	2.4, 5	74	248	OFDM	~70

Table 2.1- 802.11n standard specification [2]

2.1.2   Channel bonding

It uses a 40MHz channel which means greater bandwidth which ultimately leads to better performance in WLAN. The 40MHz channel [5] behaves like a combination of two 20 MHz channel. It can also operate in two modes: 20MHz mode and 40MHz mode. The 20MHz [5] mode is used if the bandwidth availability is limited. The 40MHz channel provides two times more desirability than 20MHz channel.

2.2   MAC layer enhancement

Frame aggregation is a feature first introduced in 802.11e Wireless LAN where two or more frame is sent in a single transmission to increase the throughput. Frame aggregation reduces overhead since it transmits multiple frames as a single aggregate frame it reduces traffic

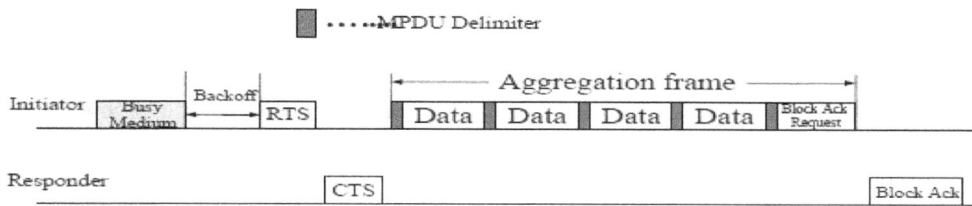

Fig 2.1- Frame Aggregation method [9]

There are three types of frame aggregation techniques- MSDU, MPDU and Two Level Frame Aggregation.

2.2.1   MAC Service Data Unit (MSDU)

This is the unit of transmission used at the MAC layer for data which is received from the upper layer. The data received from the logical link control (LLC). This LLC sub layer is present above the medium access control (MAC) sub layer in a protocol stack.

2.2.2   Aggregated MAC Service Data Unit (A-MSDU)

Multiple MAC service data unit are combined to form the payload of a MPDU packet. The entire aggregate packet will have a single frame header and these packets are destined to the same client and should have same access category. The main motivation of A-MSDU technique is the smaller size of Ethernet header when compared to 802.11 header. Hence multiple Ethernet frames can be combined to form A-MSDU. Since individual sub frames don't have frame checksum, selective retransmission of corrupted sub frames will not be possible. Sub frames will have same traffic identifier (TID) and sequence number [2]. For constructing an A-MSDU there is some limitations to follow such as, every MSDUs should have the same traffic identifier value. The sustained time of A-MSDU should be equal to the maximum sustained time of its constituent elements. In the sub frame header the values of destination address and sender address should match with the values of receiver address and the transmitter address in the MAC header.

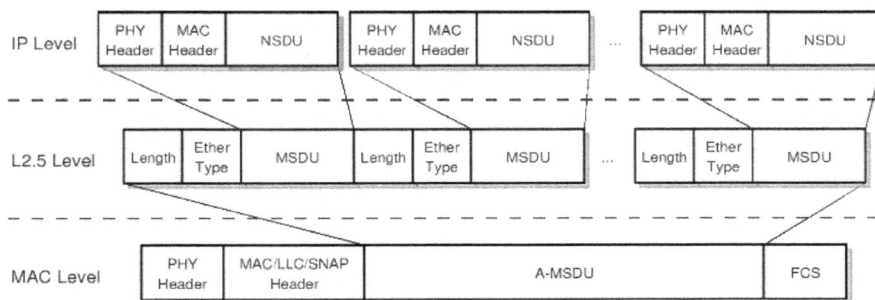

Fig 2.2- Format of A-MSDU [10]

The maximum frame length allowed in A-MSDU scheme is 3kb or 7kb [2]. Every sub frame consists of a header these headers is followed by packets that arrived from logical link control and 0 – 3 bytes of padding [10]. The padding size depends on each sub frame but the last one must be a multiple of four bytes [9].

2.2.3 MAC Protocol Data Unit (MPDU)

MAC protocol data units are the frames passed from the MAC layers into the PHY layer. This MPDU mechanism is more efficient with high data rates because of block acknowledgement. This block acknowledgement allows each of the data aggregated frames to be acknowledged individually and if error occurred it is retransmitted.

2.2.4 Aggregated MAC Protocol Data Unit (A-MPDU)

Multiple MAC protocol data unit with common PHY header are aggregated as a single A-MPDU. The size of sub frames should be multiple of 4 bytes otherwise length is adjusted by adding padding bytes [9]. Each PDU in the frame has its known header and cyclic redundancy check so in case of transmission error it is sufficient to send only the lost PDU thereby improving the throughput [3]. A-MPDU functions after the encapsulation process of MAC header here the matching traffic identifier is not considered but all the MPDUs present within an A-MPDU should be addressed to the same receiver address. Since there is no holding time to form A-MPDU the number of MPDUs to be aggregated depends on the packets that are already present in the transmission queue.

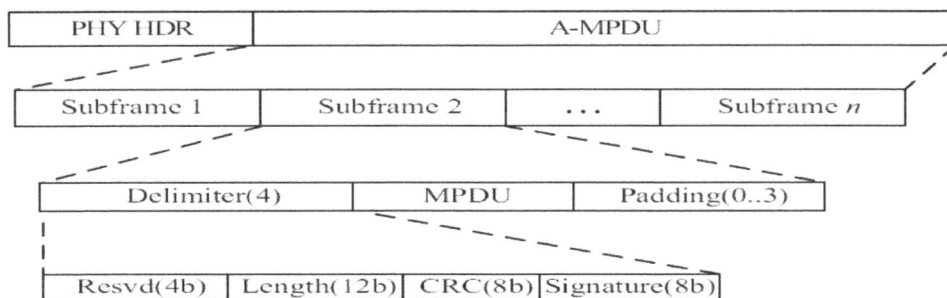

Fig 2.3- Format of A-MPDU [10]

Before MPDU a set of delimiters are added and in its tail padding bits varied from 0 – 3 bytes are added to define the length and position inside the aggregated frame the delimiter header is used. For determining the structure of A-MPDU the usage of MPDU delimiters and PAD bytes are must. After the A-MPDU is received it checks for the MPDU delimiter for any errors in cyclic redundancy check. If the value is correct then the MPDU is extracted else it continues until it reaches the end of physical layer service data unit

2.3 Block acknowledgement

The block acknowledgement is a single acknowledgement frame that acknowledges multiple received frames.802.11n allows a receiver to cumulatively acknowledge multiple frames with a single Block Acknowledgement [9].

Octets:	2	2	6	6	2	2	128	4
	Frame Control	Duration/ ID	RA	TA	BA Control	Block Ack Starting Sequence Control	Block Ack Bitmap	FCS

MAC Header

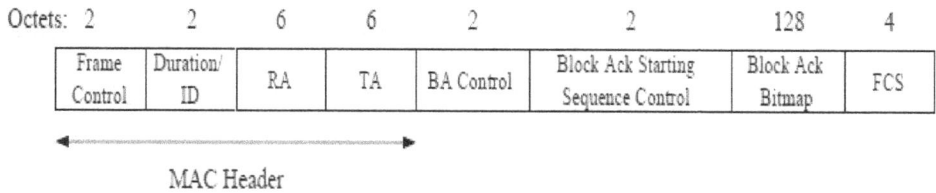

Fig 2.4 - Block acknowledgement format

To confirm the frame delivery block acknowledgment minimizes the number of ACKs that a receiver must send to a transmitter.

Access Category	Frame Aggregation	Reasons
Voice	NA	Voice packets are delay-sensitive. These packets need strict latency, jitter and bandwidth requirement [8] use of aggregation degrade the performance of voice packets since these has to wait for other packets.
Video	A-MPDU	Video packets have strict bandwidth demand but loser latency in jitter demand [8]. Video packets have high data rate so waiting for other packets exceed the maximum tolerable delay [7].
Best effort and Background	A-MPDU	Delay insensitive and no bandwidth requirements [8] making the packets of this access category will not have any impact on their performance.

Table 2.2 -Selecting required Aggregation Scheme

3.   MOTIVATION FOR THE PROPOSED WORK (Issues in frame aggregation)

Generic frame aggregation cannot be applied for all access categories because of different QoS requirements [8]. Frame aggregation is unable to function effectively in unsaturated traffic. As one side, the sender has to wait for more frames which deliberately increases the latency there by affecting delay sensitive packets. Other choice is, it is forced to send the available packet which does not use the bandwidth efficiently. In the above cases, the advantage gained by the frame aggregation is lost.

4.   PROPOSED WORK

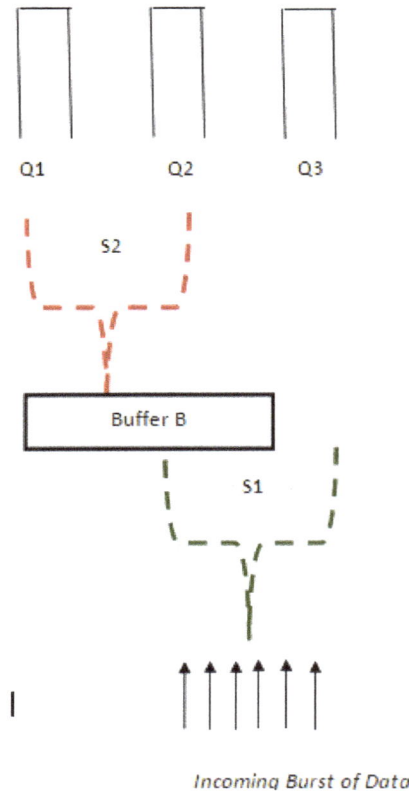

1.   **Q1**-Stores Audio packets. i.e. (Delay sensitive packets). We cannot apply frame aggregation to those packets. MSDU Concept has to be used in this case.

2.   **Q2**- Stores Video packets. We can apply frame aggregation to those packets.

3.   **Q3**- Background Best effort packets (delay insensitive packets), which can be sent anytime but will not affect the performance.

4.   **S1**-   Scheduler   1   (**Outer Scheduler**)

5.   **S2**-   Scheduler   2   (**Inner Scheduler**)

Fig 4.1- Bi-Scheduler Implementation Using Queues

4.1 Working of Bi-Scheduler Technique for Throughput improvement in Frame Aggregation

The idea is based on the fact that since Audio packets are delay sensitive and have low data rate, so making them wait in the queue will introduce lot of delay in network and degrades its performance. Thus, A-MSDU technique must be applied to such packets in order to increase their throughput. Hence frame aggregation (A-MSDU) can be effectively applied only to improve the throughput of high data rate packets. Therefore, making such types of packets wait in the queue (to attain the desired size for aggregation) does not significantly affect the throughput of the network. While Background and Best Effort packets are delay insensitive and making them wait will not have any impact in the through put of the network.

1.   Initially, the **Outer Scheduler (S1)** separates the *delay sensitive* frames i.e. Voice/ Audio frames from the incoming burst of multipart data based on their Access Category (AC). These frames are directly enqued into Queue Q1, where they wait for the other frames of same AC to arrive in order to form A-MSDU. A timer is also maintained separately for this queue in order to minimize the delay that is deliberately introduced in this process. If the timer for this queue expires, automatically the contents of Q1 will be sent to respective destination irrespective of the number of frames available at that time.

2.   The remaining frames i.e. Video and Background/ Best-Efforts are stored temporarily in Buffer B1 of **Inner Scheduler (S2)**. S2 then separates Video frames from Best-Efforts frames and enques them in queue Q2, where they are aggregated to form A-MPDU packets. The rest of the frames are enqued in Q3. A common timer is also maintained for Q2 and Q3.

   a.   When Q2 has reached the aggregated size before the timer expires, it's transferred to the respective destination.

   b.   If Q2 has not reached the aggregated size before the timer expires, then that many remaining packets from Q3 are taken to attain the desired size and sent to the destination.

3.   If, simultaneously both Q1 and Q2 have reached the desired size before the timer expires, then our algorithm gives preference to Audio packets i.e. contents of Q1 will be sent first, since they are highly delay sensitive packets. After, resetting Q1's timer, Q2 contents will be sent to the respective destination.

Fig 4.2 Bi-Scheduler Algorithm's Flowchart

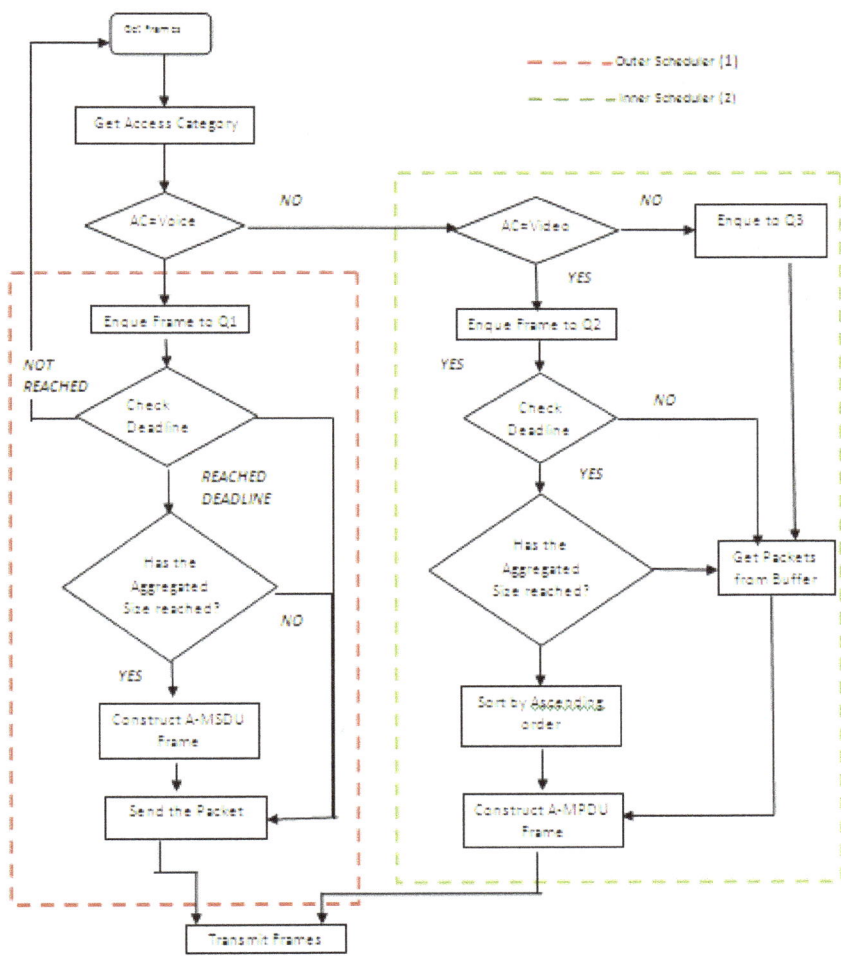

5.   CONCLUSION

The research work proposes a new Bi-Scheduler Mechanism to improve the   throughput of IEEE 802.11n WLAN using efficient frame aggregation technique. Depending on the Access Category (AC), the proposed scheduler decides whether to implement A-MSDU or A-MPDU mechanism, in order to efficiently transmit the frames to the destination. The future work will be to simulate and experimentally verify the proposed scheduler mechanism effectively.

ACKNOWLEDGEMENTS

We would like to acknowledge and thank the Defence Research Development Organisation (DRDO), New Delhi, India for granting us extra mural research funds for carrying out the research work is a part of the research project titled "Network Throughput Enhancement in distributed using Cross layer optimisation".

We would also like to thank the Head of the Department of Department of CSE and other faculty members who gave us necessary support to implement this project.

REFERENCES

[1]   Nasreddine Hajlaoui, Issam Jabri and Maher Ben Jemaa, "Experimental Performance   Evaluation and Frame Aggregation Enhancement in IEEE 802.11n WLANs".
[2]   J.Kolap, S.Krishnan, N.Shala,"Frame Aggregation Mechanism for High Throughputin 802.11n WLANS".
[3]   Selvam T, Srikanth.S, "A Frame Aggregation Scheduler for IEEE 802.11n".
[4]   WI-FI Certified TM 802.11n Draft2.0.
[5]   L.Kriara, M K. Marina, A.Farshad, "Characterization of 802.11n Wireless LAN performance via Test bed measurement and statistical Analysis".
[6]   A.Saif, M.othman, S.Subramaniam and N.A.A. Hamid "Frame Aggregation in Wireless Techniques and Issues".
[7]   S.Gubner and C. Luidemann, "Evaluating the Impact of Frame Aggregation on Video Streaming over IEEE802.11n Multi hop Network".
[8]   B.Selvig, A.Bai, T.Skeie, P. Engelstad, "Differentiated Service in 802.11e WLANS"
[9]   T.Nakajima, Y.Utsuomiya, Y. Nishibayashi, T.Tandai, T.Adachi and M. Takagi "Compressed Block Acknowledgment – an efficient selective repeat mechanism for IEEE 802.11n"
[10] Sources are also taken from Google and Wikipedia.

Why the Accuracy of the Received Signal Strengths as a Positioning Technique was not accurate?

Samir A. Elsagheer Mohamed

Electrical Engineering Department, Faculty of Engineering, South Valley University, Aswan, Egypt.
In sabbatical leave in College of Computer, Qassim University, P.O.B 6688, Buryadah 51453, KSA.
Emails: samhmd@qu.edu.sa; samirahmed@yahoo.com

ABSTRACT

There is a great need for having reliable and accurate positioning technique for many wireless network technologies such as the Wireless Ad Hoc Networks, Wireless Sensor Networks, Vehicular Ad Hoc Networks, and Wireless Local Area Networks. In addition, the Mobile Computing which still in the development phase requires good localization system for the mobile nodes. The system must allow each node to know precisely its current position or location in real time. Despite the many works that exist in the literature in this area, all of the proposals have some limitations. The features of optimal system are accurate, reliable, scalable, lightweight, and inexpensive. There are many works for the use of the Received Signal Strength (RSS) as a positioning system. However, in these works the accuracy of the positioning is very bad, about 50 meters. In this paper, we provide positioning technique based on the RSS and satisfy the mentioned optimal system features. The obtained experimental results showed an accuracy that can be better than only 2 meters. This huge improvement is due to the way by which the system is used to minimize the radio interference and the near-far problems.

KEYWORDS

Positioning Systems; Localization Systems; Neural Networks; Wireless Networks; Received Signal Stength; Curve Fitting.

1 INTRODUCTION

With the advance in the wireless networks technologies and their vast deployment, the demand on having precise localization or positioning system increases. Such system allows any wireless node (mobile device) to know its position in real time. The technologies that require such system are Wireless Local Area Networks [1], Wireless Ad-hoc Networks [2], Mobile Ad-hoc Networks [3], Wireless Sensor Networks [4], and Vehicular Ad-hoc Networks (VANET) [5-8]. Another new technology that requires a localization system is the Mobile Computing [9].

There exist several positioning techniques that have been proposed for that matter. However, the limitations in many of them make them unfeasible in practice on in wide deployment. To further detail this point, we have to know the features of the optimal system, which are accurate, reliable, scalable, lightweight, and inexpensive. No existing system can satisfy all these features at once. Lack of accuracy of the resulting measurements is the most unacceptable disadvantage. The most common positioning system in the world is the GPS [10][11] (a GPS receiver calculates its global position from the signals received from at least three satellites that feeds information about their positions). GPS has the following features: not accurate, not reliable, scalable, complex, and expensive. GPS devices can produce an error of up to 50 meters [10]. Recently an enhancement to the GPS referred as the Differential GPS (DGPS) consisting in installing expensive ground stations can improve the accuracy significantly. However, GPS and DGPS and the similar techniques do not work in tunnels, undergrounds, and in high dense building areas, because the signal cannot be received or it is received very weak. This limits the use of GPS and hence DGPS in many applications (e.g. the Sensor Networks and indoor applications). Other positioning techniques are provided in the Related Works Section.

A survey in the literature reveals that the use of the Received Signal Strength (RSS) [1][12][13][14] as a positioning technique can satisfy all the optimal features except the most important one: the accuracy. Wireless networks operate by emitting radio signals that propagate in the surroundings of the transmitter. The Signal Strength decays with the distance from the emitter. Thus, any node in the covering area of the emitter can estimate the distance to that emitter based on the Received Signal Strength (RSS). The nice features of this technique are the following. No extra hardware is required, thus it is inexpensive or free. It is scalable, as any node in the network can use it as the signals are broadcasted by the emitter and hence received by all the nodes in the covering area. It is know that the wireless networks use broadcast channels. It is reliable given that the wireless nodes are located in the covering area of the emitters. Finally, it is not complex, as simple calculation can reveal an estimate of the relative position of the node based on the RSSs. Regarding the accuracy of the RSS techniques, the existing works have one common issue: lack of accuracy (sometimes more than 50 meters as accuracy [12]).

In this Paper, we present a new methodology by which the RSS technique can provide the optimal system features. The main advantage of the proposed technique is about the accuracy. Huge improvement of the accuracy is obtained and verified by the experimental measurements. This is mainly to the way by which the radio sources are configured. That led to solve two main drawbacks: the interference and the near-far problems commonly found in the wireless networks. By minimizing the interference, the RSS gives better estimate of the distance. Furthermore, by not using the radio source signals that are very strong (when the wireless node is near to the radio source), the RSS gives a true estimate of the distance. By solving these two problems, a significant accuracy improvement is obtained. The proposed technique is very lightweight. We propose two systems: one that uses the Neural Networks (NN) and the second that uses a simple curve fitting which gives a maximum error of only 2 meters.

The rest of this paper is organized as follows: In Section 2, the related works and research efforts are given. Experimentations and data analysis that led to the discovery for improving the accuracy using RSS techniques are given in Section YYY. Descriptions of the proposed techniques are given in Section **Error! Reference source not found.**. Experimentations and the obtained results are given in Section 3. Finally, the Conclusions and the future works are given in Section 6.

2 RELATED WORKS

There exits two categories of positioning systems: indoors techniques [16][24][25][26] and outdoor techniques. We focus here on the outdoor techniques. Existing outdoor positioning techniques include the followings. The Time-Of-Arrival (TOA) [11] based techniques which are based on the travelled distance from the base station to the receiver of a known signal. TOA techniques require a perfect synchronization between the clocks of the base stations and the receivers. This cannot be guaranteed except by using atomic clocks which are very expensive. Otherwise the accuracy is compromised. The Direction-of-Arrival (DOA) or Angle-of-Arrival (AOA) techniques require multidirectional antennas and complex hardware [11]. A composed technique referred as TOA-DOA is used in [3]. Finally, there exists techniques that use the round trip time [11]. The receiver sends a small packet to the base station and wait for a reply from it. The elapsed time is proportional to the distance between it and the base station. By some calculations the receiver can know its location provided that it can talk to at least three base stations which know their accurate locations. Again the distance between the base stations and the receiver must be large to obtain good results.

The most famous one which is always mentioned in the research papers is the GPS [10][11]. However, GPS has several drawbacks as discussed below:

- Its accuracy, the civilian GPS has a limited accuracy in the order of 50 meters in each direction. A survey of the most famous GPS device vendors, the best announced accuracy is +/- 5 meters, which gives an error of about 10 meters in each direction. They say that this accuracy is obtained in 95% of all the readings. The other 5% of the readings, the accuracy may be very bad. This still too much far away of being acceptable in the critical applications requiring a precise positioning system.

- In some places (e.g. inside tunnels, undergrounds, etc.), the GPS has no coverage. This is because the GPS receivers calculate the position based on the signals received from 4 simultaneous satellites. In such places, the received signals are too weak or no signals can be received at all. Therefore, the GPS receivers cannot calculate the position.

- GPS signals sent from the satellites can be simulated. In other words, there exist GPS simulators (devices generating fake GPS signals). As mentioned before, the real GPS signals are relatively weak signals. GPS simulators produce relatively strong signals and therefore, the GPS receiver will use the fake signals and consider the real signals background noise. As a result, the receiver will calculate the position based on the fake signals. Thus, any jammer equipped by this kind of GPS simulator can inject erroneous data that leads to make the vehicle believe that it is in completely different position.

Depending on the wireless network technology, there are several studies proposing several poisoning techniques. Examples are: Wireless Local Area Networks [1][27], Wireless Ad-hoc Networks [2], Mobile Ad-hoc Networks [3], Wireless Sensor Networks [4], Vehicular Ad-hoc Networks (VANET) [11] and Mobile Computing [9].

3 FIELD EXPERIMENTS TO MEASURE RSSS FROM WAPS.

In order to develop and validate the positioning system based on the RSS, two field experiments are conducted in realistic conditions. In wireless network configurations, the IEEE 802.11g standard is used. The network architecture consists of having Wireless Access Points (WAPs) installed in fixed and know locations. A Mobile Unit (MU) which is simply a Laptop having the software that can measure the RSS from the WAPs at any place in the covering area is used. In both experiments, the WAPs are installed in a straight line and the distance between each two successive WAPs is 100m. The MU moves in a line parallel to the WAPs line and it measures the RSSs from all the WAPs at predefined location. We have used these field experiments' data for a positioning technique for the Vehicular Ad-Hoc Networks [15].

3.1 The first experiment

The environment and the precautions that we take for this experiment are as follows. The MU measures the RSSs each 5m from the starting point (location 0m) until the ending point (location 200m). The elevation of the access points is 110 cm, above the ground. Distance between the line of access points and the parallel line of the test line equals 7m. All the access points and mobile host operates on Channel 6. A diagram of the setup of this experiment is shown in Figure 1.

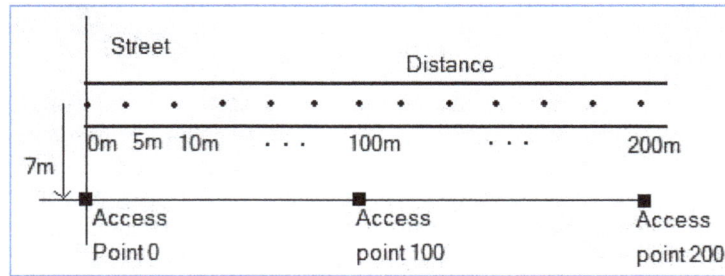

Figure 1. Experiment #1 Setup

3.1.1   Obtained Results for Experiment #1

In this Section, we will present the obtained results. In all the Figures in this Section, the x-axis shows the distance starting from the 0m location. The y-axis shows the value of the received signal strength intensity (RSS) from the wireless access point(s) WAPs. Remember, the RSS is negative and the far we are from the WAP, the less the RSS.

Figure 2   shows the obtained RSS values from the three Access Points for the First Experiment (Experiment #1). From the shown data, we can see that the results are very bad, the data is very chaotic. It is expected that the RSS from any WAP decrease when we move far away of it (increasing the distance). However, for all the WAPs, this is not the case. By inspecting the results, we can expect that no way to obtain accurate estimate of the position based on the RSSs. On other words, the accuracy will be very bad. That expectation is confirmed using the neural network as a learning tool (See Section 4.1.2). The accuracy is very bad, about 80 meters in any direction (for the lack of space, results are omitted).

A very interesting finding here is that this obtained result by the neural network (NN) is almost the same as those found in the literature [12][11]. It seems that the existing work in the RSS techniques do not take into consideration the effect of the channel interference.

3.1.2   Why the results are chaotic?

That was the first experiment that we carried out. Once we collected the results, we have thought in the problem. We, then, realized that all the WAPs operated on the same frequency channel. It is known that the Wi-Fi networks have 11-13 channels depending on the country. If all the WAPs use the same channel, that will cause significant noise and interference. Another problem is that this experiment setup suffers from the near-far problem described before.

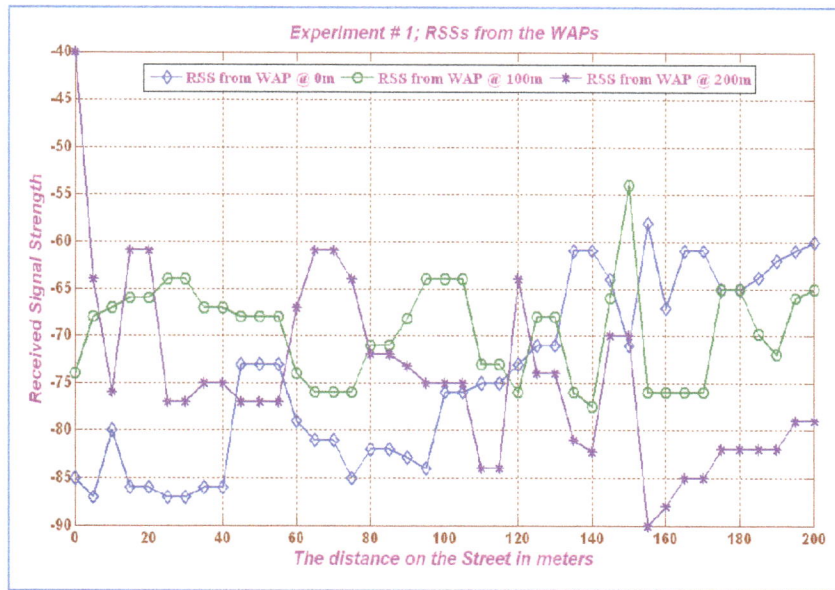

Figure 2. Obtained RSS from the three Access Points for the First Experiment (Experiment #1)

3.2   The Second Field Experiment

In this experiment our goal is to solve a fatal problem discovered in Experiment #1. It is known that the channels in IEEE 802.11 standards do not overlap if they are more than 5 channels apart. For example, if one WAP use channel #1, the second uses channel #6 and the third uses channel #11, then there is not overlap and thus no noise nor interference that can degrade the results. That is confirmed using the obtained results. Thus in this experiment, to reduce the radio channel interference and hence increase the accuracy of the RSSI, each WAP operates on a non-overlapping channel (Chanel #1, 7, and 13). All other experiment setup is the same as the previous one. We have repeated the experiment three times with different configurations. We obtained the same results in each case. Thus, in this Section, only one of the results will be presented and analyzed.

3.2.1   Experiment #2 obtained results

Figure 3 depicts the obtained RSS values from the three Access Points for Experiment #2.

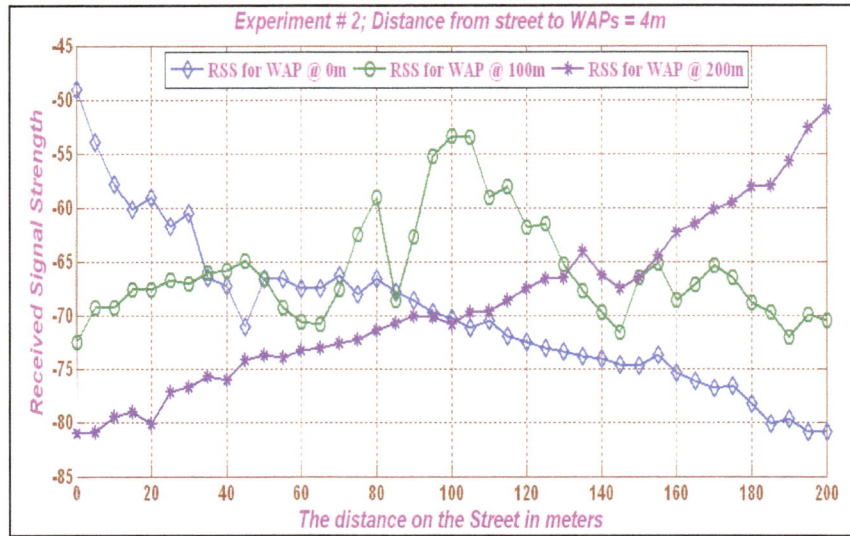

Figure 3. Obtained RSS from the three Access Points for the Second Experiment (Experiment #2).

3.2.2   Data Analysis of Experiment #2

As can be seen from Figure 3, once the interference of the channels is eliminated, the results are improved significantly. Moreover, one of the most important discoveries in this work is that the RSS from any WAP is still chaotic even if there is no interference if the distance to the WAP is less than 60 meters. When the MU moves away from the WAP, we obtain better results (better estimate of the distance based on the RSS). This can be seen for WAP@0m and WAP@200m. When the distance is greater than 60 meters, the RSS is inversely proportional to the distance (the expected behavior). For WAP@100m, this is also true. In the middle part of Figure 3, the data are chaotic. However, it starts to be correct at the boundaries.

In the next Sections we present the use of the neural networks and the curve fitting to estimate the position of the vehicle based on the received signal strengths and the use of the obtained experimental data.

4 TWO COMPETITIVE POSITIONING SYSTEMS USING RSS

Based on the results of the presented field experiments and the data analysis, two positioning systems are presented in this Section. First, we explore the use of the Neural Networks and we analysis their performance and limitations. Second, a simple, but efficient technique using the curve fitting is proposed.

4.1   A positioning system using the Neural Networks

The use of the neural networks (NNs) as a learning tool to estimate the location of the MU based on the RSS from the WAPs is presented in this Section. The full theory and the description of the NN are beyond the scope of this paper. However, the basic functionality of the neural networks is given here. The idea of the NN is inspired from the biological nervosystem. Similarly to the biological neuron, the artificial neural network learns by experience. In other words, a set of learning examples must be collected for the problem in hand. These samples are simply a set of the system inputs and their corresponding outputs. Using training algorithms, the NN will learn the system behavior based on these examples. Once trained, the neural network will act as the system. For any new inputs, the NN will estimate the outputs based on the learning experience.

4.1.1 Description of the approach with the NN

Matlab 2009a is used to do most of the calculations, and the Figures in this paper. Especially, the powerful toolbox in Matlab called the 'Neural Network toolbox' is used for all the work in this Section.

The inputs to the NN are the RSS from the WAPs. The output is the location of the MU in the field. The NN needs to be trained, validated then tested. Thus, using Matlab, the whole learning example (the set of samples in the form RSS-Position pairs), is divided into three sets: one for training; the second for validation; and the third for testing. This is done automatically by Matlab. The ratios are 70% for training; 15% for validation; and 15% for testing. The training stops when the performance on the validation set increase. This is to avoid a known problem with the NN which is the overtraining.

For the NN, we use the Feedforward architecture that contains the input layer connected to the inputs; the output layer, which gives the estimate of the location based on the learning experience. In addition, there is the hidden layer. This layer can contain a given number of hidden neurons. Moreover, the random seed for the random generator which is used to assign the initial weights of the NN is very important. For the same samples, and the same NN, but different random seeds, the trained networks are completely different. Thus, we take this as a parameter for all the data.

A Matlab script to automate most of the work is developed. Then the outputs can be saved and restored at any time to know the best NN that gives the best performance.

4.1.2 Working with data obtained in experiment #1

We have run the script on the data obtained from experiment #1. A total of 56 different NN architectures are created; trained; tested and validated. The performance measures are also calculated on both the testing and the whole samples.

Figure 4 shows the difference between the actual data (distance) and those estimated by the trained NN for the best one. As we can see from the Figure, error on the positioning may reach up to 60m. This is very high and not suitable for most the wireless networks. This bad result is that found on the literature, see for example [11]. Actually, people used the technique without taking into account the effect of the interference problem resulting from using the same channel for all the WAPs. This is why the published results for the mechanisms using the RSSs are not encouraging.

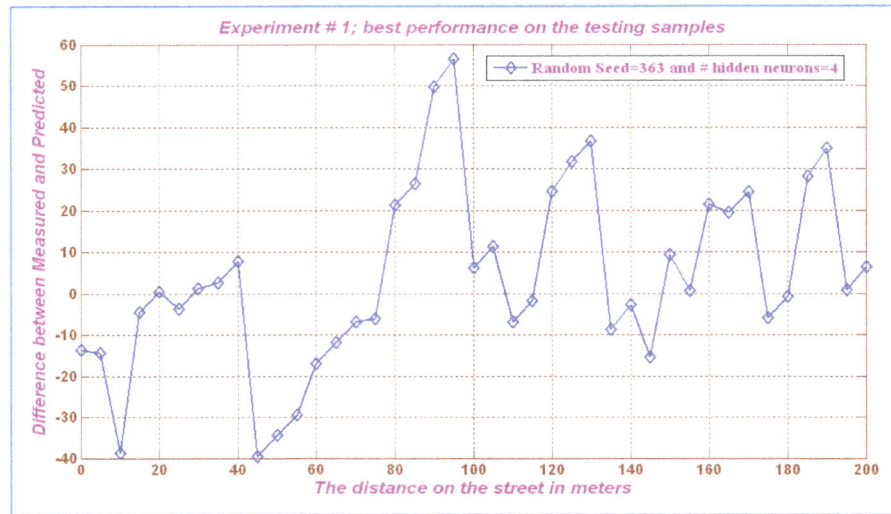

Figure 4. The performance of the best NN for experiment #1. The Figure shows the difference between the measured data experimentally and those calculated by the NN. The error can reach up to 60 meters. Correlation coefficients= 93 %

4.1.3 Working with data of experiment #2

Similar to what done in experiment #1, the whole work is repeated on the data for experiment #2. We have run the script on the data obtained from experiment #2. By using NNs with different hidden neurons and random seed generator: a total of 180 different NNs are trained, validated and tested. The performance measures are also calculated on both the testing and the whole samples. In Table 1, we show the obtained data sorted by the MSE on the whole samples and the max error obtained. In the table, we show also the standard deviation, the max error, the variance and the correlation for both the testing samples and the whole samples.

From this table, we can see that the best NN is obtained when the # of hidden neurons is 10 and for initial random seed =102. As expected the accuracy of this best one is very good compared to that of the first experiment. The MSE of the best NN is 6 against 172 for the first experiment. The standard deviation is about 3.2m, the max error is about 7.6 meters against 60 meters. Comparing these results to the GPS, we can see that we can reach accuracy better than the GPS. Thus, our technique can be used to complement the GPS in the areas where the GPS signal cannot be received or if it is very week.

Thus, the accuracy is improved significantly, by about a factor of 7. That is a huge improvement. This can also be seen from Figure 5 which depicts the difference between the actual data (distance) and those estimated by the trained NN.

4.1.4 Effect of removing the WAP@100m

We have removed the RSS from WAP@100m that we have thought may degrade the performance. The results show no further improvement (omitted for the sake of space). This is because three WAPs are better than two.

Table 1. Results of exp. #2. By using NNs with different hidden neurons and random seed generator: A total of 180 different NNs are trained, validated and tested; only the top 5 are shown

Row #	# Hidden Neurons	Random Seed	Mean Square Error (MSE)		Max Absolute Error		Standard Deviation		Variance		Correlation Coefficient	
			Testing Items	All Items	Testing Items	All Items	Testing Items	All Items	Testing Items	All Items	Testing Items	All Items
1	10	102	8.4	6.0	5.4	7.6	3.2	2.5	10.0	6.1	0.997	0.999
2	8	102	15.9	7.5	8.0	8.0	4.2	2.7	17.4	7.2	0.997	0.999
3	8	1	5.9	7.9	3.5	8.8	2.4	2.8	5.9	8.1	1.000	0.999
4	4	899	8.2	8.1	5.4	8.6	2.9	2.9	8.5	8.3	0.999	0.999
5	9	363	38.6	8.3	12.4	12.4	5.5	2.8	30.1	8.1	0.998	0.999

Figure 5. The performance of the best NN for experiment #2. The Figure shows the difference between the measured data experimentally and those calculated by the NN.

4.1.5   Using only one WAP but far away

We are seeking to improve the performance as much as it possible. Thus, in this subsection, we will use data from WAP @0m and WAP@200m only when the distance from each one is greater than 60m. This is to eliminate the source of error due to the chaotic part of the curve (near-far problem). The obtained results also are not encouraging. Again, the NN cannot correctly learn from the few samples available for training. The next subsection presents another way to improve the performance, which is the curve fitting.

4.2   Positioning System Using Curve Fitting

In the previous Section, we have seen how to use the neural network for the positioning system. However, it is still in the range of 7 meters. We are seeking to improve this performance. We have explored the use of the curve fitting to obtain a polynomial equation that can be used to obtain the distance from the WAP given its RSS. Several type of curve fittings are evaluated using Matlab. The best results is given bellow. We aim to approximate the positioning problem according to the following equation. For the Linear model Polynomial of degree 4, we have the following equation:

$$\text{Distance} = p_1 R^4 + p_2 R^3 + p_3 R^2 + p_4 R + p_5,$$

where R is the received signal strength.

4.2.1   Starting from 60m

We have used the data from WAP@200m and by removing the data when the distance is less than 60m (the chaotic data). Using Matlab "cftool" curve fitting tool, the estimated values of the Coefficients (with 95% confidence bounds) are as follows:

p_1= -0.005206, p_2= -1.553, p_3= -173.5, p_4= -8608, p_5= -1.601e+005.

The obtained Goodness of fit for the above data is: SSE: 422.7; R-square: 0.9917;   Adjusted R-square: 0.9903; and RMSE: 4.197. The curve fitting results are shown in Figure 6. We have calculated the distance for the given RSSs in the range. In Figure 7, we draw the difference between the actual data (measured) and the calculated ones. One can see that using only one WAP, the accuracy is almost the same as using the NN with three WAP. That is a very important finding.

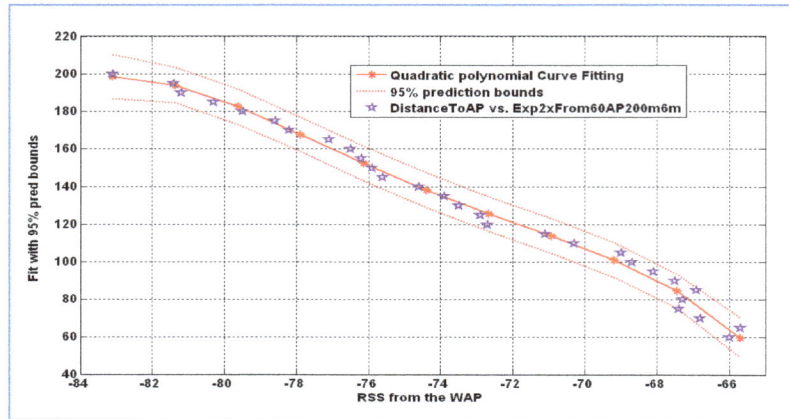

Figure 6. Curve Fitting results using cftool, when the Distance is greater than 60m

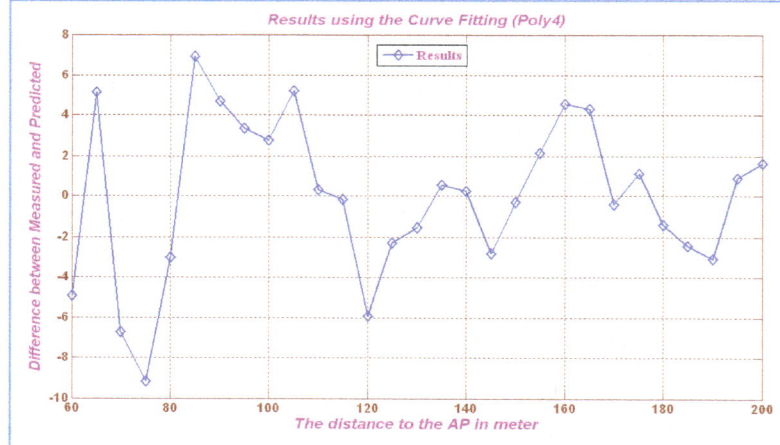

Figure 7 The performance curve fitting for Exp # by using only data from WAP@200m and excluding data before 60m. The Figure shows the difference between the measured data experimentally and those calculated by the NN.

4.2.2   Starting from 100m (solving the near-far problem)

From the previous section, we can see that the performance is bad at the beginning of the curve. However, it is getting better when the distance from the WAP increase. Thus, we redid the same experiment, but removing all the data when the distance is less than 100m. The results of the curve fitting are shown in Figure 8.

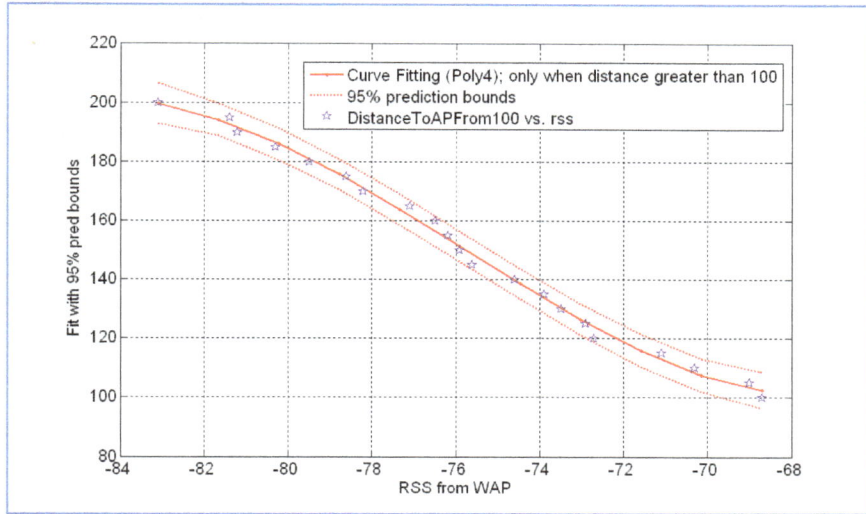

Figure 8. Curve Fitting results using cftool, when the Distance is greater than 100m

The coefficients of the polynomial equation are as follows: p_1=0.0004373, p_2=0.1736, p_3=24.4, p_4=1458, p_5 =3.171e+004. The obtained goodness of fit is: SSE=85.13, R-square=0.9956, Adjusted R-square=0.9945, RMSE= 2.307. Comparing these data with the ones in the previous experiment, we note a very significant improvement. This is also confirmed from Figure 9. We can see that the accuracy is improved (the max error is now 4 meters instead of 8 meters). This is by using only one WAP. Again, if we inspect the figure carefully, we can see that the results when the distance is greater than 170m are within only 2 meters (A very accurate results). This result is much better that that obtained using the traditional GPS.

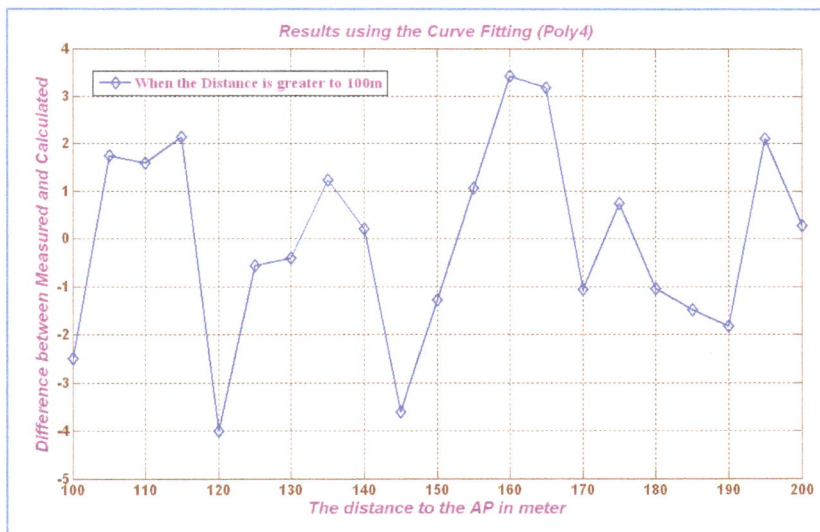

Figure 9. The performance curve fitting for Exp #2 using only data from WAP@200m and excluding data before 100m. The Figure shows the difference between the measured data experimentally and those calculated by the NN.

5 THE BEST APPROACH FOR THE POSITIONING SYSTEM

Based on all the results explained in the previous Sections, we summarize here the best approach that can be used to build a system for an accurate positioning system using the RSSs. The resultant system gives accurate positioning result because the interference and the near-far problems are solved. In the wireless network geographical area, there must be several radio sources. For example, for the WLAN the wireless access points (WAPs) are the radio sources that periodically emit beacons. Second example, in the WSN, several reference WAPs have to be installed. Third example, in VANET, the Road-Side Unites will act as the WAP.

- The WAPs must be installed with, if possible, a minimum distance between each two of 200m. They have to cover the whole geographical area.

- At any point, the mobile unite (MU) unit must receive from at least two WAPs satisfying the next condition: these WAPs must be as far as possible (the less the RSS), and each operates on a different radio channel.

- In order to be able to obtain the position of the MU, it must hear from at least two WAPs, if the 2D location of the MU is desired. In the case of 3D positioning, at least three access points must be heard.

- If the MU hears from many WAPs, this will be very good. This can improve dramatically the accuracy. For example, in order to obtain the 2D position of the MU, only two WAPs are need. But if it receives from three, we find three readings. By taking the average of the three values, the accuracy will be improved. Adding one WAP can improve the 3D case very much.

- Using a simple triangulation technique, the position coordinate point of the MU can be determined.

- Each WAP knows its absolute location. Thus, the MU can determine its absolute location based on the estimated relative position and the absolute locations of the surrounding WAPs.

6 THE CONCLUSIONS AND FUTURE DIRECTIONS

In this paper, an accurate generic positioning technique for the wireless networks is proposed. It is based on the Received Signal Strength (RSS). Although, the use of the RSS is not new, the way by which it is used in this paper to significantly improve the accuracy is new. The improvement of accuracy is due to tackling the problems of the channel interference and the near-far problems. The obtained results are an accuracy of 2 meters compared to 50 meters in most of the published works.

Field experiments are conducted on realistic situations. In the first one, the WAPs operate on the same channel. Hence, the interference caused degradation of estimating the distance from the RSS. In the second experiment, the results are very good compared to the first as the WAPs operated on different channels to reduce the interference. After analyzing the obtained results, we have proposed the use of the neural networks (NN) to learn the positioning problem from the obtained learning examples. By using the curve fitting technique, and using only one WAP, we have done two experiments that show an obtained accuracy that can reach a maximum absolute error of only 2 meters. This can be obtained when the WAP is very far from the mobile unit. Future research directions include the exploration of the technique in the specific wireless technologies like the Wireless Sensor Networks.

REFERENCES

[1] C. Komar, and C. Ersoy, "Location Tracking and Location Based Service Using IEEE 802.11 WLAN Infrastructure," in *Proc. of the European Wireless Workshop*, February 2004.

[2] Misra, S.; Guoliang Xue; Bhardwaj, S.; Secure and Robust Localization in a Wireless Ad Hoc Environment. IEEE Transactions on Vehicular Technology, Volume: 58, Issue: 3 2009, pp: 1480 – 1489.

[3] Zhonghai Wang and Zekavat, S.A., "A new TOA-DOA node localization for mobile ad-hoc networks: Achieving high performance and low complexity" IEEE 17th International Conference on Telecommunications (ICT), 2010, pp. 836 – 842

[4] Weng Kai and Chen Chun; "Using RSS with difference method in localization algorithm for sensor networks", 2nd International Conference on Information Science and Engineering (ICISE), 2010; pp. 2500 – 2502

[5] K. Saha and D. B. Johnson, "Modeling mobility for vehicular ad hoc networks," in *Proc. of the ACM Workshop on Vehicular Ad hoc Networks (VANET)*, October 2004, pp. 91–92.

[6] R. Tatchikou, and S. Biswas, "Vehicle-to-vehicle packet forwarding protocols for cooperative collision avoidance", in Proc. of Globecom 2005, St. Louis, USA, Nov. 2005.

[7] A.Nandan, S. Das, G. Pau, M.Y. Sanadidi, and M. Gerla, "Cooperative downloading in Vehicular Ad Hoc Networks", in Proc. of WONS 2005, St. Moritz, Switzerland, Jan. 2005.

[8] S. Kato, S. Tsugawa, K. Tokuda, T. Matsui, and H. Fujii, "Vehicle control algorithms for cooperative driving with automated vehicles and intervehicle communications," *IEEE Trans. on Intelligent Transportation Systems*, vol. 3, no. 3, pp. 155–161, September 2002.

[9] Vaughan-Nichols, S.J. Will Mobile Computing's Future Be Location, Location, Location? Computer Journal, Volume: 42 , Issue: 2, DOI: 10.1109/MC.2009.65, 2009 , pp. 14 – 17

[10] Yu, K.; Guo, Y.J.; Anchor Global Position Accuracy Enhancement Based on Data Fusion. IEEE Transactions on Vehicular Technology, Vol.: 58, Issue: 3, 2009, pp.: 1616 – 1623

[11] M. Porretta, P. Nepa, G. Manara, and F. Gianne, "Location, Location, Location", IEEE Vehicular Technology Magazine, Vol. 3; pp. 20-29, June 2008

[12] Bo-Chieh Liu; Ken-Huang Lin; SSSD-Based Mobile Positioning: On the Accuracy Improvement Issues in Distance and Location Estimations. IEEE Transactions on Vehicular Technology, Volume: 58, Issue: 3. 2009, Page(s): 1245 – 1254.

[13] Ouyang, R. W.; Wong, A. K.-S.; Lea, C.-T.; Received Signal Strength-Based Wireless Localization via Semidefinite Programming: Noncooperative and Cooperative Schemes. IEEE Transactions on Vehicular Technology, Volume: 59, Issue:3, 2010, pp: 1307 – 1318

[14] Parker, R.; Valaee, S.; Vehicular Node Localization Using Received-Signal-Strength Indicator. IEEE Transactions on Vehicular Technology, Volume: 56, Issue: 6, Part: 1 2007, pp: 3371 – 3380

[15] S.A.Mohamed, A. Nasr, G.A. Ansari, "Precise Positioning Systems for Vehicular Ad-hoc Networks using Received Signal Strength and Carbon Nanotube", submitted to Computer Journal, Oxford.

[16] Hossain, A.K.M.; Hien Nguyen Van; Yunye Jin; Wee-Seng Soh; "Indoor Localization Using Multiple Wireless Technologies", In IEEE Internatonal Conference on Mobile Adhoc and Sensor Systems, 2007. MASS 2007. pp: 1 – 8.

[17] Rahman, M.Z.; Kleeman, L.; Paired Measurement Localization: A Robust Approach for Wireless Localization IEEE Transactions on Mobile Computing, Vol.: 8, Issue: 8, 2009, pp: 1087 – 1102

[18] R. Singh, M. Guainazzo, and C. S. Regazzoni, "Location determination using WLAN in conjunction with GPS network," in *Proc. of the Vehicular Technology Conference (VTC-Spring)*, May 2004, pp. 2695–2699

[19] Bo-Chieh Liu; Lin, K.-H.; Jieh-Chian Wu; Analysis of hyperbolic and circular positioning algorithms using stationary signal-strength-difference measurements in wireless communications. IEEE Transactions on Vehicular Technology, Volume: 55, Issue: 2, 2006, pp: 499 – 509.

[20] Mada, K.K.; Hsiao-Chun Wu; Iyengar, S.S.; Efficient and Robust EM Algorithm for Multiple Wideband Source Localization. IEEE Transactions on Vehicular Technology, Volume: 58, Issue: 6, 2009, pp: 3071 - 3075

[21] X. Chen, Y. Chen, and F. Rao, "An Efficient Spatial Publish/Subscribe System for Intelligent Location-Based Services", Proceedings of the 2nd International Workshop on Distributed Event-Based Systems (DEBS'03), June 2003

[22] J. Hightower and G. Borriello, "Location Systems for Ubiquitous Computing," Computer, vol. 34, no. 8, pp. 57-66, IEEE Computer Society Press, August 2001

[23] L. Wischof, A. Ebner, and H. Rohling, "Information dissemination in self-organizing intervehicle networks", IEEE Trans. intelligent transportation systems, vol. 6, no. 1, pp. 90-101, Mar. 2005.

[24] Saad, S.S.; Nakad, Z.S.; , "A Standalone RFID Indoor Positioning System Using Passive Tags," IEEE Transactions on Industrial Electronics, vol.58, no.5, pp.1961-1970, May 2011 doi: 10.1109/TIE.2010.2055774

[25] Chen, Miao; Jin, Jishun; , "An Improvement Approach of Indoor Location Based on Radio-Map Using Wireless Local Area Network," 2011 International Conference on Computer and Management (CAMAN), pp.1-4, 19-21 May 2011, doi: 10.1109/CAMAN.2011.5778740

[26] Ming, Wang; Zhengqiu, Lu; Shan, Jin; Beng, Shao; Qingzhang, Chen; , "Design of an indoor positioning and tracking algorithm for wireless sensor networks," 2011 International Conference on Consumer Electronics, Communications and networks (CECNet), pp.3081-3084, 16-18 April 2011doi: 10.1109/CECNET.2011.5768290

[27] Shum, K.C.Y.; Quan Jia Cheng; Ng, J.K.Y.; Ng, D.; , "A Signal Strength Based Location Estimation Algorithm within a Wireless Network," 2011 IEEE International Conference on Advanced Information Networking and Applications (AINA), pp.509-516, 22-25 March 2011doi: 10.1109/AINA.2011.80

11

ANALYSIS OF CELL PHONE USAGE USING CORRELATION TECHNIQUES

DR. T S R MURTHY[1] AND D. SIVA RAMA KRISHNA[2]

[1]Professor,Shri Vishnu Engg. College for Women,Bhimavaram. West Godavari dist,A.P,India
e-mail: sriram_pavan123@yahoo.co.in

[2]Asst. professor,Swarnandhra Institute of Engg. & Tech.,Narsapuram. West Godavari dist, A.P,India
e-mail: srkd999@gmail.com

ABSTRACT

The present paper is a sample survey analysis, examined based on correlation techniques. The usage of mobile phones is clearly almost un-avoidable these days and as such the authors have made a systematic survey through a well prepared questionnaire on making use of mobile phones to the maximum extent. These samples are various economical groups across a population of over one-lakh people. The results are scientifically categorized and interpreted to match the ground reality.

KEY WORDS

Correlation coefficient, Coefficient of Determination, Probable Error, Standard Error of Correlation coefficient.

1. INTRODUCTION

With the advent of communication technology, the whole world has come under one umbrella. Next to internet, cell phone has changed the life style of people and business persons in particular. The cell phone has become an indispensable tool for one and all. Right from a school going child, a house wife to a servant, the cell phone has its major impact on their lives.

Several studies have shown that mobile phone in various communities has positive as well as negative influence on Society. The cell phone is the need of every one. It is for the fact that now- a- days, having a mobile phone is a sort of necessity and it is an inevitable truth that the mobile industry is taking every one by the storm. In spite of its advantages, there are also some disadvantages of mobile phone. So it all depends on the user how he makes use of it for better living.

The Following objectives have been formulated in order to quantify various aspects that influence the use of mobile phones
.

- Correlation between the male and female using mobile phones.
- Correlation between the age groups of various people using the mobile phone.
- Correlation between the utilization of mobile phone with regard to the occupation of all sectors of people.
- Correlation between the maximum mobile usages of various brands by maximum number of people.

- Correlation between the maximum purchase of mobile phones with respect to the influence made by friends, neighbors, advertisements, cell cost and similar reasons.
- Correlation between the users suffers from the battery backup.
- Correlation between the study with regard to various networks (SIM cards) utilized by the mobile users.
- Correlation among different networks (SIM cards) whether they influences the purchaser.
- Correlation between the study of net work connection with respect to pre-paid and post paid connections.
- Correlation between the monthly expenditures.
- The relation between the correlation and probable error
- Relation between the correlation and 6 times of probable error

Methodology

To satisfy the objectives of the study, qualitative methodologies along with quantitative techniques are employed. The study is in descriptive nature. A total of 2131 people were made to participate in this survey. Only 1804 of these people are using mobile phones and the remaining 327 are non-users of cell phones because of their own reasons. The survey is made, keeping in mind all classes of people such as business persons, farmers, house wives, workers, students, employees and educationalists. The survey is also made, keeping in mind different economic groups like, people of low class, middle class and high class. All the survey is made in the town of Bhimavaram as it has all types of people and business practices. The survey is made in eight different localities of Bhimavaram to meet the over all objectives of the present work.

The Primary Data was collected by the MBA students of 2009-11 Batch of Shri Vishnu Engineering College for Women, Bhimavaram. Out of 2131 people, nearly 500 high class people, 1000 middle class people and 500 low class people in the society were taken into consideration. With the help of the faculty and the students of MBA, a questionnaire is framed after detailed discussions to get holistic understanding of subject. In this paper the correlation between the eight localities with respect to the above said objectives are measured.

The Secondary Data is collected from various magazines, news papers and internet websites on different aspects of mobile phone utilization. After collecting the data from various sources, the data is subjected to verification, quantification and coding with referred coding keys. Then, the coded data is keyed into computer for data processing and analysis. The Statistical packages R, SPSS, Microsoft-Excel are used for calculation, percentage, frequency, distribution,

Results and Analysis

Correlation is a statistical tool for studying the relationship between two or more variables. Theory of correlation deals with the observation and measurement of the relationship between two or more statistical series. The study of correlation is very important from business and economic point of view.

It helps in verifying the reliability and accuracy of the data that analyses cause and effect relationship. Correlation does not necessarily mean causational. Even in the case of significant correlation, there may be no cause-and-affect connection. If a change in one variable causes a change in the other variable, is called correlation between the two variables. A change in one may be caused by or due to the other but the statistical evidence may not indicate which is cause and which is affect. Correlation may be due entirely to a third factor or several other factors affecting each variable, this is the case in time series. Correlation analysis involves various methods and the techniques used for studying and measuring the extent relationship between the two variables.

The figures show the joint variation among the pairs of values and give an idea of the relation ship between the two variables. If the points are scattered around a straight line, the correlation is linear and if the points are scattered around a curve, the correlation is non-linear, if the points are scattered all over without any pattern, there may not be any correlation between the variables. Two variables may have a positive correlation or negative correlation or they may be uncorrelated.

It may be noted that the all the points representing the pairs of values lie around a straight line with a positive slope. This indicates that the correlation is positive and is linear. The value of correlation coefficient lies between -1 and +1. The value +1 implies that there is a perfect positive correlation, and the pairs of values of x and y lie on the straight line with positive slope. The value -1 implies that there is a perfect negative correlation, and the pairs of values of x and y lie on the straight line with negative slope.

However, in real life situations, when the variables are random variables, the correlation coefficient does not equal the value 1 (+1 or -1). Its value nearing 1, indicates strong linear relationship and its value nearing 0 indicates absence of linear relationship.

Thus the zero value of correlation coefficient does not imply that there is no relationship between the two variables or they are independent of each other. However, if two variables are independent of each other then the correlation coefficient is zero.
There is interesting feature about interpretation of the low/high values of correlation and the extent of linear relationship.

In general, to have confidence in the interpretation of the value of correlation, one should have a good number of observations.

Significant relationship exists only when correlation coefficient is greater than 0.7, when it is greater than or equal to 0.9 then the relationship between the two variables is highly significant, since coefficient of determinant is greater than or equal to 0.81. If the correlation coefficient is less than the probable error then there is no evidence of relationship between the variables (i.e., there is no significant relationship between the variables). If the correlation coefficient is greater than six times the probable error then it is considered as a significant relationship between the variables.

- Regarding the mobile phone users, the correlation between the males and females is 0.86, shown that there is a very high degree of correlation between the males and female.

- With respect to age groups concern, the age groups 20-40 and 40-60 show a highest degree of correlation of mobile users which is 0.89. and the next highest is 0.79 between the age groups less than 20 and 20-40. The least is found with less than 20 years and greater than 60 years age group. It is shown in the table 1, and from figure1, we observe that the correlation between the age group 20-40 and 40-60 is highly significant and there is a positive correlation between them.

- Regarding the occupation, the correlation between the business people and employees is 0.84; this is also a very high degree of correlation between the business men and employees.

- With regard to mobile companies, correlation between various brands with a target brand *A* is measured, since brand *A* is the leading of all the remaining brands as observed from the table 1. In this, correlation between brand *A* and brand *B* is the highest degree of correlation which is 0.93. From figure 2, it is observe that the correlation between the mobiles *A* and *B* is highly significant i.e., the two variables are correspondingly change.

- It can also observed that the correlation between brand *A* and brand *F* is 0.88 which is the next highest correlation, and the lowest correlation between brand *A* and brand *C* is 0.33. i.e., the correlation between the mobiles *A* and *C* is very poor, this shows in figure 3, there is a change in one variable then the corresponding change in the other variable is very slow. This may be due to several reasons, the brand *C* may not be lucrative, may not provide more service options, may not possess more features. As such, it cannot reach more public.

- Regarding the brands of mobiles *A&B*; *A&E*; *A&G*, the relation between correlation and probable error; correlation and 6 times probable error are found significant.

- In sales point of view, Correlation between Advertisement and cell cost is 0.84 that shows a high degree of association between the advertisements and cell cost. And the correlation between the cell cost and others is 0.71. From these two, it is observed that cell cost plays a major influence in sales of the mobile phone.

- Regarding the battery back-up of cell phone, the correlation between the Average and Good is 0.63, which is also a high degree of correlation, and the lowest correlation is 0.21 between Average and Poor battery backup.

- Regarding sales of network, it is observed from the table 1, the brand a is taking a leading role when compared to the remaining brands, only with a percentage of about 36. Correlation between various brands with brand *a* is measured, for brand *a* is the leading of all the remaining brands. In this, correlation between brand *a* and brand *b* is the highest degree of correlation which is 0.93. From figure 4, it is observe that, even though there is an evidence of correlation between a and b, for some values of a, the other variable b maintains silence. And the next highest is correlation between brand *a* and brand *c* is 0.92, and for remaining brands with brand *a* is also high. And the lowest of all these with brand *a* is 0.42 the correlation between *a* and *i*. From figure 5, it is observe that the correlation between the networks *a* and *i* is poor, i.e., if the change in one variable there is a poor change in the other variable. The low correlation with brand *i* network may be attributed to its less coverage and weak signal strength.

- It has been observed that from table 1, the relation between (regarding the variables *a* and *b*) correlation and probable error; correlation and 6 times of probable error is significant. Also observed that from table1, the correlation and probable error; correlation and 6 times of probable error the relation is insignificant.

- Considering the sales of network, the correlation between advertisement and call cost is 0.41 which is a low degree of correlation between them. And the correlation between call cost and others is 0.59 which is a high degree of correlation. In this the call cost plays a major role in choosing the networks by the users and this is shown in the table 1. The maximum number of users purchased cell phones at their own discretion.

- The correlation between the prepaid and post paid connections is 0.83.The majority of mobile users are pre-paid customers with a percentage of nearly 80.

- Finally, regarding monthly expenditure of the mobile users, the highest degree of correlation between 'less than Rs.300' and 'Rs.300-600 customers' is 0.96 from the figure 6, it is observed that there is a corresponding change between the two variables. and the lowest degree of all these, the correlation between Rs.300-600 and Rs 600-1200 is 0.71.

- It is evident that from table 1, the relation between (regarding the variables <300 and 300-600; <300 and >600) correlation and probable error; correlation and 6 times of probable error is significant. Also it is observed that from table1, the correlation and probable error; correlation and 6 times of probable error the relation is insignificant regarding 300-600 and 600-1200.

Table 1.

Category										Correlation Coefficient	S.E (r)	P.E(r)	6.P.E	Remarks
Gender														
Male	295	117	94	140	125	167	181	147	1266	0.865	0.089	0.06	0.36	Significant
Female	208	107	79	120	79	78	119	75	865					
Age										C(<20&20-40,>60)				
<20	80	20	18	18	33	36	36	35	276		0.134	0.09	0.54	Significant
20-40	223	127	53	60	60	136	184	140	983	0.7887	0.277	0.187	1.123	Insignificant
>60	11	8	7	6	4	7	11	13	67	0.4644	0.071	0.48	0.287	Significant
20-40	223	127	53	60	60	136	184	140	983	C(20-40&40-60)				
40-60	105	65	37	52	37	49	69	64	478	0.8939	0.071	0.048	0.287	Significant
Occupation														
Business	156	66	51	43	41	79	80	67	583	0.8348				
employees	100	48	41	53	40	73	89	42	486		0.107	0.072	0.433	Significant
F/W	57	34	7	33	16	38	37	25	247	0.4964	0.266	0.179	1.078	Insignificant
others	106	72	16	7	37	38	94	118	488					
Mobile company										C(A&B,C,D,F,G)				
A	244	115	61	71	76	142	182	135	1026	0.9269				
B	43	25	8	12	12	29	25	30	184		0.049	0.033	0.201	Significant
C	14	9	12	12	9	6	14	8	84	0.333	0.314	0.212	1.2	Insignificant
D	33	28	13	18	14	12	21	17	156	0.6781	0.191	0.129	0.774	Insignificant
E	25	17	9	6	6	9	9	11	92	0.7405	0.159	0.107	0.645	Significant
F	11	6	2	2	3	14	3	11	52	0.5789	0.234	0.158	0.95	Insignificant
G	49	20	10	15	14	16	46	40	210	0.8768	0.081	0.055	0.33	Significant
Influenced Factors										C(Mob Cost & Advert, others)				
Mobile cost	108	28	1	14	14	62	20	17	264					
Advertise	94	34	15	24	13	42	61	34	272	0.844	0.101	0.068	0.412	Significant
Others	172	104	82	78	98	94	86	140	854	0.718	0.101	0.067	0.407	Significant
Battery Back up										C(Advt&good,poor)				
Average	63	70	10	19	22	42	40	58	324					
Good	328	140	85	109	99	173	238	176	1348	0.631	0.212	0.143	0.861	Insignificant
Poor	28	10	20	8	13	13	22	18	132	0.217	0.336	0.227	1.363	Insignificant
Networks										C(a&b,c,d,e,f,g,h,i)				

a	151	69	41	51	42	89	115	80	638					
b	34	10	9	9	9	20	17	18	126	0.926	0.05	0.034	0.204	Significant
c	66	50	16	19	18	48	53	42	312	0.915	0.057	0.038	0.233	Significant
d	41	30	18	16	18	11	27	23	184	0.715	0.172	0.116	0.699	Significant
e	25	14	9	16	12	13	14	15	118	0.791	0.132	0.089	0.536	Significant
f	36	17	7	6	6	13	8	19	112	0.773	0.142	0.095	0.575	Significant
g	8	0	2	2	2	0	3	4	21	0.679	0.19	0.128	0.768	Insignificant
h	47	23	10	16	19	28	60	43	246	0.84	0.104	0.07	0.422	Significant
i	11	7	3	1	8	6	3	8	47	0.478	0.272	0.184	1.104	Insignificant
Influenced Factor										C(Advt& Cc,Others)				
Advertise	20	27	14	9	8	17	10	23	128					
Call Cost	197	54	17	19	48	131	84	108	658	0.4049	0.295	0.199	1.197	Insignificant
Others	127	97	68	88	76	65	86	101	708	0.5934	0.229	0.154	0.927	Insignificant
Connections														
Prepaid	305	192	97	106	102	189	230	218	1439					
Postpaid	114	28	18	30	32	39	70	34	365	0.8319	0.108	0.073	0.44	Significant
Monthly Expenditure										C(<300&300-600,>1200)				
<300	184	119	52	64	88	144	113	52	816					
300-600	114	70	42	45	58	75	75	25	504	0.9639	0.025	0.016	0.101	Significant
>1200	48	8	18	6	25	31	28	6	170	0.8029	0.125	0.084	0.509	Significant
300-600	114	70	42	45	58	75	75	25	504	C(300-600&600-1200)				
600-1200	73	23	24	19	57	50	36	32	170	0.7089	0.176	0.118	0.712	Insignificant

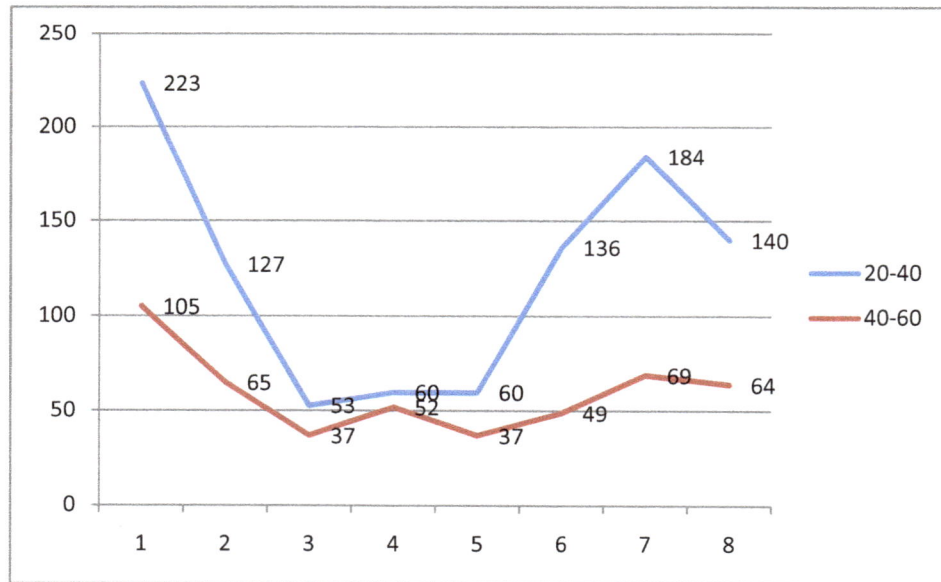

Figure 1: The correlation between the age groups 20-40 and 40-60

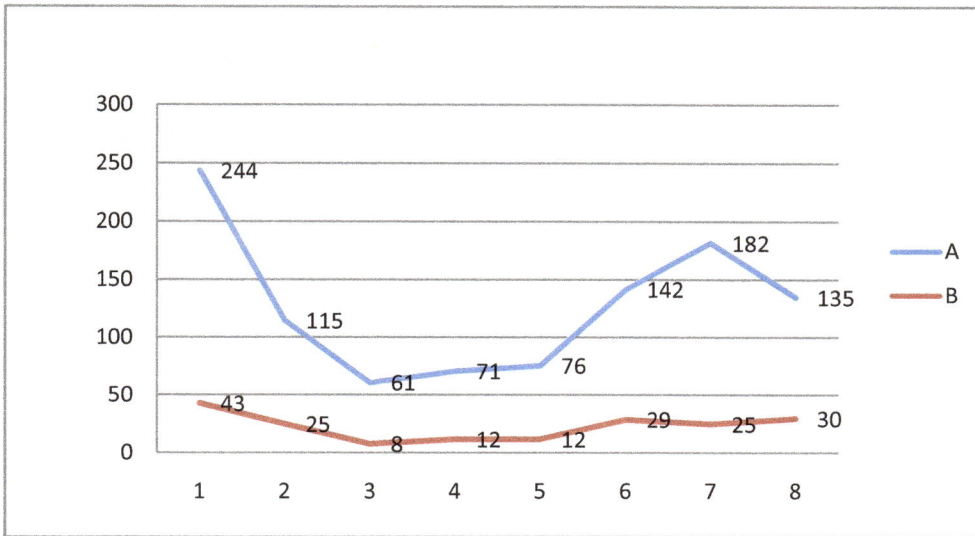

Figure 2: The correlation between the mobile companies *A* and *B*

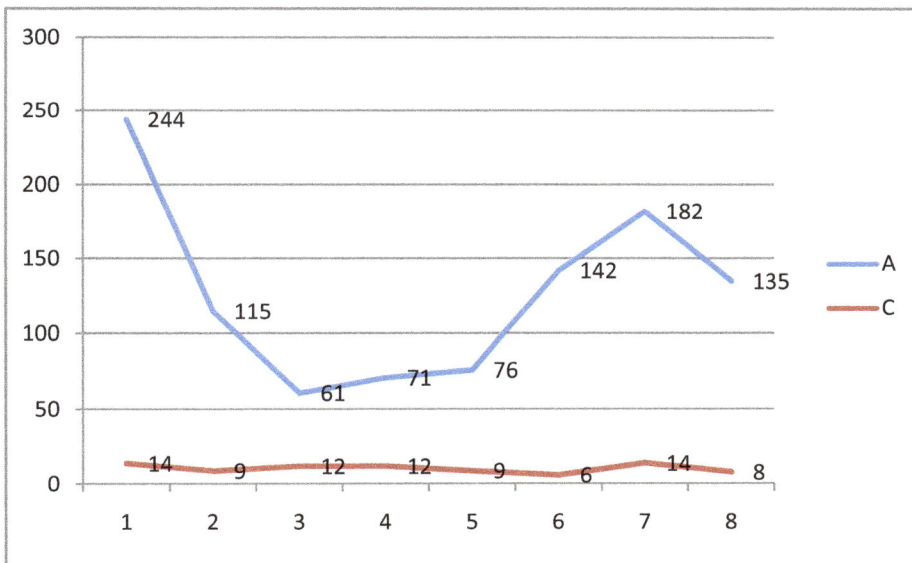

Figure 3: The correlation between the mobile companies *A* and *C*

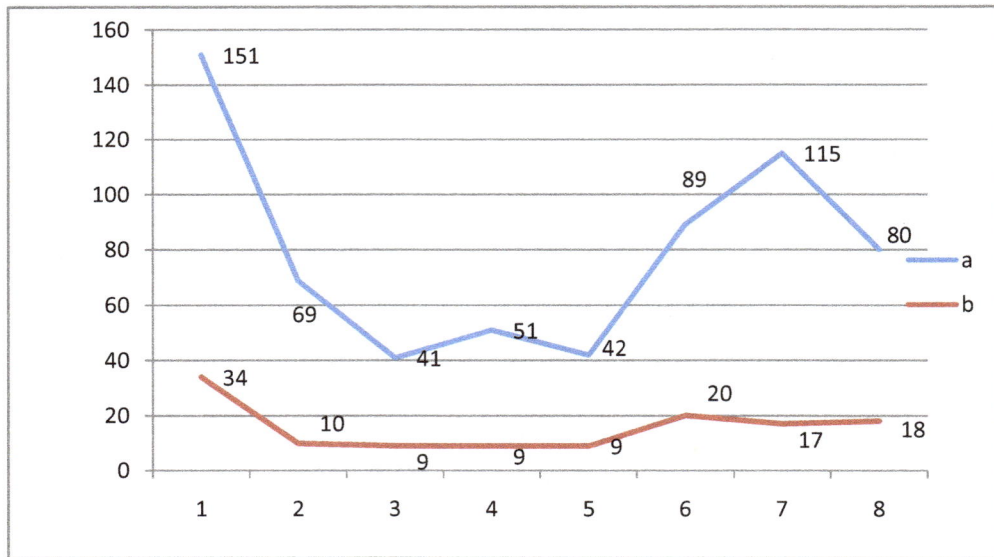

Figure 4: The correlation between networks *a* and *b*

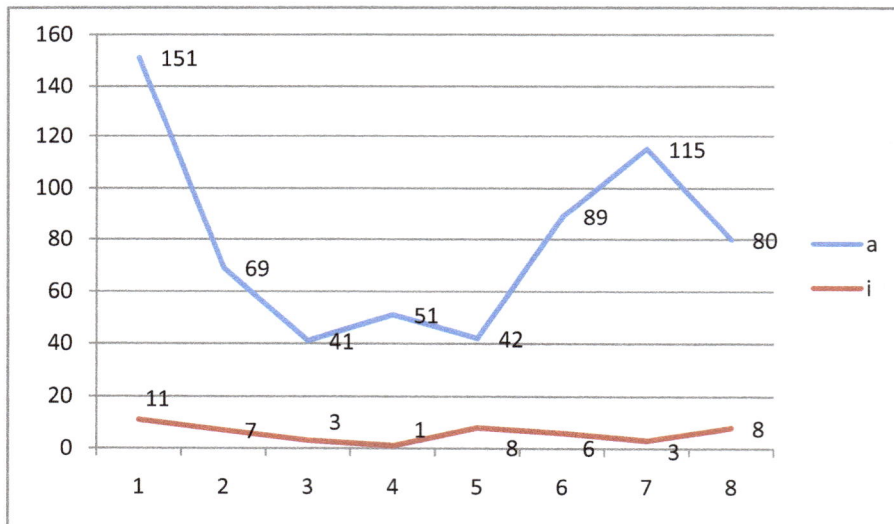

Figure 5: The correlation between the networks *a* and *i*

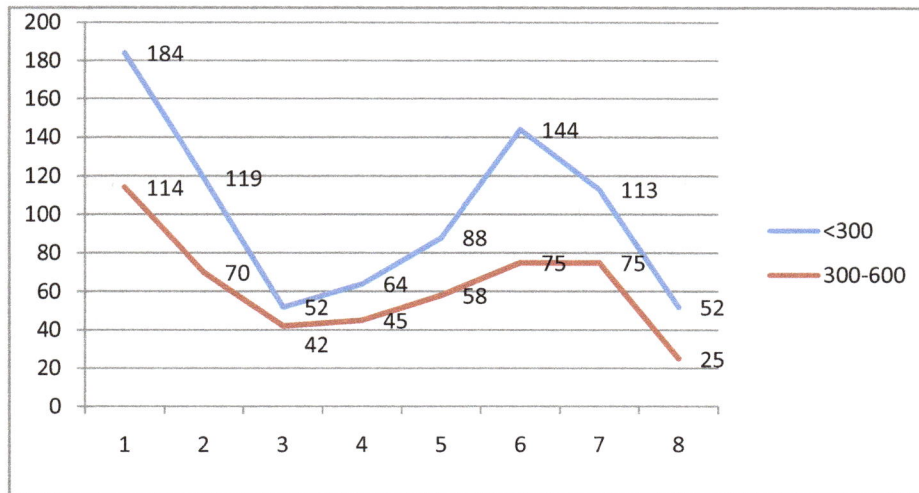

Figure 6: The correlation between the monthly expenditure Rs < 300 and Rs.300-600

Conclusions

The study of correlation is of immense use in practical life. Correlation measures the degree or strength of the linear relationship between the variables. With the help of correlation analysis one can identify the direction or nature of relationship between variables. Probable error of the correlation coefficient is applicable for the measurement of reliability of the computed value of the correlation coefficient. If the correlation coefficient is greater than six times the probable error then it is considered as a significant relationship between the variables. With this analysis, we can measure the magnitude or degree of relationship existing between the variables. Cell phone is being widely used for many valid reasons. The usage of it varies from one age group to the other. Investigation has proved that majority of users are men. Business persons and employees are also using the cell phones more than the others. There is only one mobile company 'Brand-*A*' which is selling the maximum number of cell phones and is far ahead of others. The people who are buying it are very intelligent and taking their own decisions in the choice of cell phones. The other factors like friends, neighbors, cell cost and advertisements are less influencing factors in the purchase of cell phones. The people that have been taken into account are both mobile users and non-mobile users. Out of this number, a negligible or a very few percentage of people is not using the cell phone for their own reasons.

People are also selecting network connections with different options in mind. Prepaid connection has influenced the most mobile users. Majority users have optimized with regard to the usage of cell phone. The minimum expenditure is less than Rs.300/- per month.   Only less percentage of the users spends more than Rs.1200/- per month.

Though the study is restricted to a particular town the samples are known to interviewers closely.   Another serious study is required with larger sample size so that important factors other than these may be found which influence the society to use mobiles so frequently.

ACKNOWLEDGEMENTS

The authors are thankful to the students of Shri Vishnu Engineering College for Women, 2009-2011 batch of MBA for their cooperation in collecting the data, and also thankful to their managements to provide necessary facilities.

REFERENCES

[1]   Gangadhar B.Sonar and Santosh M.Jainapur(2006)-"Impact of Mobile   Phone usage on academic Environment:A Study from Management Perspective".(Vol.2No.3 Oct 2006 Issue of Journal cf Global Economy, Mumbai).

[2]   Fraunholz B and unnithan C (2004) critical success factors in mobile communication:A comparative roadmap for germony and india international journal of mobile communication,Vol.2.,No.1.pp.87-101

[3]   Merisavo M,Vesanen J, Arpponen Aet al (2006),"The Effectiveness of Targeted Mobile Advertising in Selling Mobile Services: An Empirical Study International Journel of Mobile Communications,Vol-4,No.2,PP.119-127.

[4]   Singh Fulbag and Sharma Reema(2007),"Cellular Services and Consumer Buying Behaviour in Amritsar City ,The IUP Journel of Consumer Behaviour ,Vol.2,No.3,PP.39-51.

[5]   D. Estrin, R. Govindan, J. Heidemann, and S. Kumar, "Next century challenges: Scalable coordination   in sensor networks", in Proceedings of ACM Mobicom, Seattle, Washington, USA, August 1999, pp 263-- 270, ACM.

[6]   I.F. Akyildiz, W. Su, Y. Sankarasubramaniam and E. Cayirci, "A Survey on Sensor Networks", IEEE

        Communications Magazine, pp. 102--114, August 2002.

[7]   D. Estrin, L. Girod, G. Pottie, M. Srivastava, "Instrumenting the world with wireless sensor

        networks", In Proceedings of the International Conference on Acoustics, Speech and Signal Processing (ICASSP 2001.

[8]   L. M. S. D. Souza, H. Vogt and M. Beigl, "A survey on fault tolerance in wireless sensor networks",2007.

9]   T.N. Srivastava, Shailaja Rego " Statistics for Management" – Tata Mc Graw-Hill

[10]   Barry Render, Ralph M. Stair, JR, Michael. E. Hanna " Quantitative analysis for Management – Pearson education, 9th edition.

12

IMPLEMENTATION OF APPLICATION FOR HUGE DATA FILE TRANSFER

Taner Arsan, Fatih Günay and Elif Kaya

Department of Computer Engineering, Kadir Has University, Istanbul, Turkey

ABSTRACT

Nowadays big data transfers make people's life difficult. During the big data transfer, people waste so much time. Big data pool grows everyday by sharing data. People prefer to keep their backups at the cloud systems rather than their computers. Furthermore considering the safety of cloud systems, people prefer to keep their data at the cloud systems instead of their computers. When backups getting too much size, their data transfer becomes nearly impossible. It is obligated to transfer data with various algorithms for moving data from one place to another. These algorithms constituted for transferring data faster and safer. In this Project, an application has been developed to transfer of the huge files. Test results show its efficiency and success.

KEYWORDS

Network Protocols, Resource Management in Networks, Internet and Web Applications, Network Based Applications.

1. INTRODUCTION

Data means raw fact collection from which conclusions may be drawn. Letters of handwritten, printed books, photographs of the families, movie on video tape, printed and signed copies of mortgage papers, bank's ledgers and account holder's passbook are examples of the data.
In the past, people use the limited forms to save or share their data for example; paper or film. Then, computer invented and they can convert the same data into variety forms by the computer such as an e-mail message, text documents, images, software programs or etc. Then, these all datum are stored in strings of 0s and 1s in a computer. Since, all computer data is binary format and this forms data is called digital data as shown in Figure 1.

Data is classified of two types which are structured and unstructured as shown in Figure 2. Structured data uses row and columns format to definite and order the information. This data is stored in database management system (DBMS). Unstructured data is not stored in row and column. Data is stored in variety forms such as e-mail, images, pdf, or etc. Unstructured data is more than structured data because of do not require the storage space [1].

From past to present, information increased even it came to present exponentially. Result of this there is one expression occurred is called "Information Trash". A lot of software companies worked about this topic and Big Data expression formed. Big data is a form that data which is collected from community media sharing, network dailies, blogs, photographs, videos, log files and formed meaningful manner. According to some belief which is destroyed now data which is not structural was worthless, but big data showed us that there would be occurred how important, useful data from the manner is called Information Trash. Big Data is consist of web server logs,

statistics of internet, social media publications, blogs, microblogs, climate sensor and information which is coming from similar sensors, Call Records from GSM operators.
Big Data allows users and companies to manage risks better and making innovation if it used with true analysis methods on the right way.

Most companies are still making decisions through result of their conventional data store and data mining. But to be able to foresee consumer wishes, there should be big data analysis and to be able to proceed according to these analysis. Big data analysis, flows, store is difficult to deal with using old traditional database tools and algorithms. Computers and data storing unit is not enough for big data and not enough have capacity to use. By 2012 data is produced 2.5 quintillion bytes daily in the today's world. Handling, transferring this big data or issues about it is called Big Data.

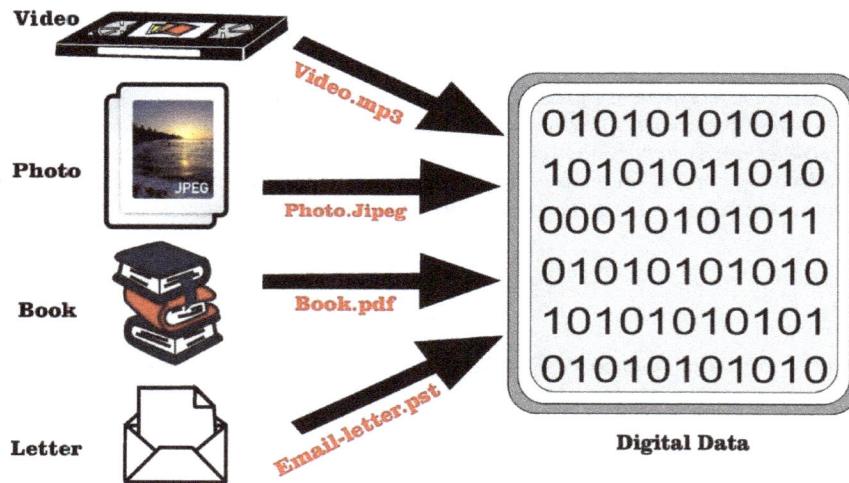

Figure 1. Digital Data

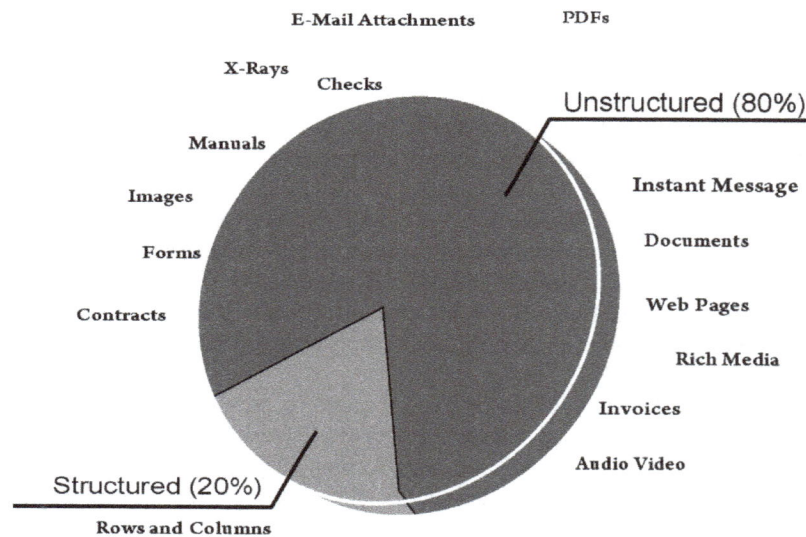

Figure 2. Data Types: Structured and Unstructured.

Databases are not enough to hold this growing data today. When relational database can hold at gigabyte level, with big data it could store at petabyte level. But big data is only proper for Batch calculations. There is no property which is critical in developed database like Transactions. Because of the reading, writing and updating databases could make through transactions, these processes accepted as atomic process and changing from different processes and making it inconsistent can be blocked. Big data should be used when it is written in once and read many times. Because data is parallel processed more than one place. This big data is produced in many sectors like RFID sensors, social media, hospitals etc. Notably data which are come from DNA sequence analysis, weather forecast sensor as a big data is need for us in many areas like data process areas.

There are 5 components in big data platform which are variety, velocity, volume, verification and value. Generally it is called 5V.

Variety: Data which is produced not structure %80 and every new technology can produce data in different formats. Its needed to deal with data type in variety comes from phones, tablets, integrated circuit. They are also in different languages, could be Non-Unicode, they need to be translated and formed in same format.

Velocity: Producing time big data is too fast and it has been growing. Process number and variety of related to the data is increasing too.

Volume: According to IDC Statistics data amount will be 44 times bigger than 2009's data amount in 2020. The fact that we called it huge and big systems which we use will be bigger 44 times than now. This fiction has to made that how to deal with data store, proceed, integration technologies with this much data volume. In 2010's total amount of informatics expense was increased by the ratio of % 5 but, amount of data producing was increased %40.

Verification: There is also another component that must be exist is security in data density. During the flow, it must be seen by true people, controlling with enough security level, and must be latency.

Value: The most important component is creating value. Big Data has to create positive value for company after the data producing and processing as describes as at the above. For example governmental institution which makes decisions and strategies about health must see illness, cure, and doctor distribution in district, city, and county etc. details. Air Forces must see every inventory's location and situations and must watch histories about them. A bank should see the person's information including eating, vacation habits and even social media activities for giving the credit to that person.

Our daily life needs and applications for them on the Internet, after sales records for the customer delight and storing data in the companies for this reason, it caused problem about lack of free space. It was the beginning of the new search about this area. Recently companies have to keep their customer's data with new services as individual and personal.

Hospitals; keeping data of their patients in a numerical environment for serve them individually and efficiently.

Governments; have to keep data of their citizens and have to work on it, for example according to rules of governmental authorities, media records have to be stored for last 1 year.

Because of the producing and consuming data on the internet affected internet service providers about ensuring big data forming meaning and reuse again.

Banks became more efficient about internet banking, more updated about daily records of the customers, availability for 7 days 24 hours online.

Power supply companies have to record of individual data by using smart system and counter. Pharmaceutical Industries have to be available always for researches and big genomic databases such as cancer researches.

Additionally GPS, GSM and cameras which can produce high resolution pictures and sounds data decreases efficiency of data storage areas. Social media platforms like Facebook and Twitter push entrepreneurs through the big data because of the data which have to be stored.
Entrepreneurs, investors and media-consulting companies have opportunities about Big Data. Clouding technology has spread around and became cheaper so that its usage increased and also it changed balance of data forming economy. After the evolution of most important technology market on Big Data, its expecting that market will exceed 50 billion dollar in 5 years. Annual increase of data in the world now is 59% and it's expecting to increase more. Traditional and new data sources are laying on this growth. IDC digital records will be 1.2M zeta bytes at the end of the year and it will be 44 times more than now in this following 10 years. Essential source of growth is on structured data. Myth about 80% of unstructured data is worthless is now destroyed because of the result of the success for e-commerce companies and search motors. Main necessity is storing of structured and unstructured data, and analysing them and data mining.

In 1980s companies' main aim was the produce main object and provide to access to customer when production was more important. Essential purpose for the developing of ERP systems is collecting customer, distributing centres, supplier and production in one platform. People started to ask this question "who is the right customer for me?" when this system has enough satisfaction.

Born of CRM systems also started with the same question, CRM focus one point which is "providing right production to the right customer with right price by right way on right time and in right place." So now it's not customer for the production, its production for the customer. This methodology keep increase its importance in last 10 years.

2. BIG DATA COMPUTING

Thanks to the developments of communications, digital sensors, computation and storage have generate very big collections of data, and catching information of prize for the business, science, government and society. Most popular search engine companies such as Google, Yahoo! and Microsoft have developed a new sector and business making the information freely accessible on the World Wide Web and support to the users in functional ways. These companies gather trillions of bytes of information every day and they are developing their services such as satellite images, driving directions, and image retrieval. These services are advantages of sociality such as collecting people's information, gathering them together, knowing how to use people's data and their value are measureless.

Some of the big data computing forms will change the companies' actions, scientific researches, practitioner of medical, security intelligence and nation's defence like how search engines have changed accessing of the data for people.

In the modern medicine world, patient's information is collected via some technological equipment. Such as CAT scans, MRI, DNA microarrays, and the many other of equipment. Due to collection of big data sets of patients and applied them to data mining, many developments are possible for the medical world.

The collection of data on the World Wide Web is showing up to be a principal and can be mined and ran in various ways. For instance language translation programs can be used by statistical language models which are created by result of analysis of the billions of documents in the languages which we want to translate from and the source, along multilingual documents. Advanced web browsers spare documents according to the reader levels of English from the beginners to adults. Carnegie Mellon University has a research of information storage on the human brains and in which way human brains are storing the information. The study depends on word combinations which are existing in the web documents.

The big data technology has the importance has growing because of the many technologies.
Sensors: Many different kind of sources creating digital data such as digital imagers, chemical and biological sensors, and foundations and people. Such as digital cameras, MRI machines, microarrays, environmental monitors.

Computer Networks: Localized sensor networks such as Internet can gather data together from the various sources.

Data Storage: Thanks to the magnetic disk technology development, cost of storing data has decreased. For instance cost of storing one trillion bytes of data in a one terabyte disk drive is about $100.Acoording to the reference of the one estimation which is converting the all books in the Library of Congress to the digital form can be done it with around 20 terabytes .

High Speed Networking: In spite of the fact that one terabyte data can be stored on a disk for $100, it's transferring with a group is requires one hour or more and approximately a day by "high speed" Internet connection. Out of the ordinary the most common way to transfer mass data between sites is to guide disk drive through Federal Express. Bandwidth limitations expanding the difficulty of efficiency usage of computing and storing resources in a group. The power of connecting a cluster and end user which are geographically distributed also limited by them. The difference between amounts of data which is using for storing and which is using for communication will increase. For the combination of increasing bandwidth and decreasing cost, there should be "Moore's Law" technology.

Cluster Computer Programming: Programming for the large scale of the data in distributed computer systems is an existing problem since running large sets of data become required in an optimum time. The software should divide the data and make computations between nodes in a data set, and whenever there is an error occurs in the hardware or software, software must investigate and fix the error. Great inventions such as MapReduce programming framework presented by Google, these inventions are built for the organization and programming the systems. To understand the power of the big data computing, more effective and powerful techniques must be improved.

Expanding the access of cloud computing: AWS has a technological limitations such as bandwidth, and it's not appropriate for the large data and its computations, however Amazon making good profit with AWS. Additionally, limited bandwidth causes time consuming and expenses for taking data inside and sending it out. The cloud systems should be separated from each other because the fact that decreasing the sensibility for the disasters. But data mobility and interoperability is needed in more developed levels. There is a project called The Open Cirrus which is directing attention to this point by creating international environment to make experiments on connected cluster systems. Organizations must tune to new costing model on their managerial part. For example, governments do not take money from the universities for the capital costs such as purchasing new machines but they do for operating costs. We can imagine that ecology of cloud systems which are supporting general capacity of computing and right on target special services or storing expert data sets.

Security and Privacy: Data sets include sensitive data and tools which can be used for take-off information and produce opportunities for access without authorization and use. Many of privacy protections in our community are based on current uselessness. For instance people have been watching by video cameras from many locations such as markets, ATMs, airport security. The possibility of abuse becomes important when these sources gathered together and advanced technology of computing let it associate and analyzing the data strems. For harmful agents, cloud facilities became a cost effective platform for the start a botnet or implement huge parallelism for breaking the cryptosystem. We must generate assurances to prevent misuse together with evolving technology to enable beneficial properties.

Remarkably companies which have available environment for the Internet service comes first in the industry. These companies also providing billions of dollars for their computing systems and overshadow the conventional ones. Companies like Google, Yahoo!, and Amazon are the pioneers of the cluster computing systems' installation and programming. Other companies receiving news of the business benefits and the operating efficiency which are discovered by these companies.

University researchers do not have enough access to large scale cluster computing and new sight's appreciation. This situation is quickly changing because of their colleagues' success in the industry, providing access and training. Google, IBM, Yahoo! and Amazon ensure access of their computing resources for the students and researchers. It is enough to make satisfy the some potential needs but clearly it's not enough for providing all potential needs for spreading the application of data intensive computing. Some of the large scale scientific projects are making plans for managing and providing computing capacity for their stored data. There is a spirited debate between new approximation of data management and traditional one. Weakness of research funding is an important block for attendance and encouraging of university researches.
The development of big data computing is one of the best innovations in recently. The importance of its properties such as collecting, organizing and processing of the data has been recognized newly. It's simply to provide the development of it by the federal government's financial support.

3. CONCEPT OF DATA TRANSFER

Data is transferred by some applications such as electronic mail, file transfer, web documents, so bandwidth and timing are important things for data transfer. Figure 3 refers data transfer;

Figure 3. Data Transfer

If you want to transmit small data, you need small rate bandwidth such as the application of internet telephony encodes voice at 32 kbps. However, if you have huge files and want to transmit them, you need more bandwidth. This is more advantages than small rate bandwidth.
Timing is important when you transmit the data. Applications should provide quick data transferring to save time. For example, real-time applications of internet telephony, virtual environments, multiplayer games or etc.

3.1. Internet Transport Protocols

Two transport protocols are used to applications which are UDP and TCP [2]. Each of these protocols provide different service model for applications, so you should choose one of them when you create a network applications for the Internet [3].

3.1.1. Transmission Control Protocol (TCP)

Transmission Control Protocol (TCP) guarantees the reliable data transfer. TCP service models include connection-oriented service as shown in Figure 4. The connection is a full-duplex between two hosts. TCP exchanges data between applications as a stream of bytes. Also, TCP implements highly congestion-control mechanism. TCP has high quality connection between initiator and receiver as shown in Figure 5.

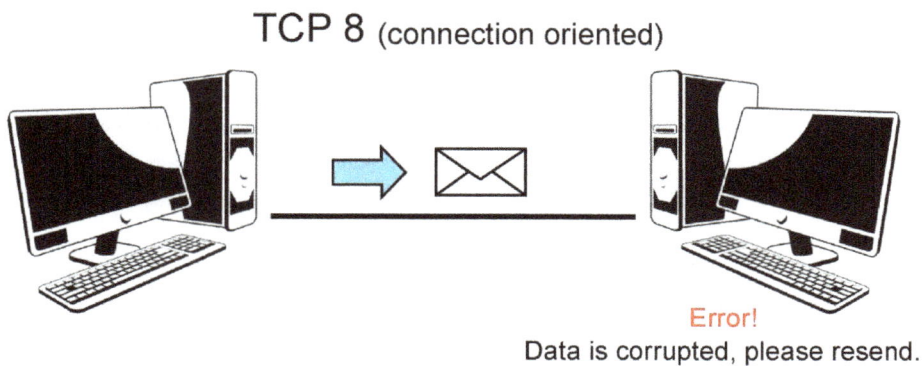

Figure 4. TCP (connection-oriented)

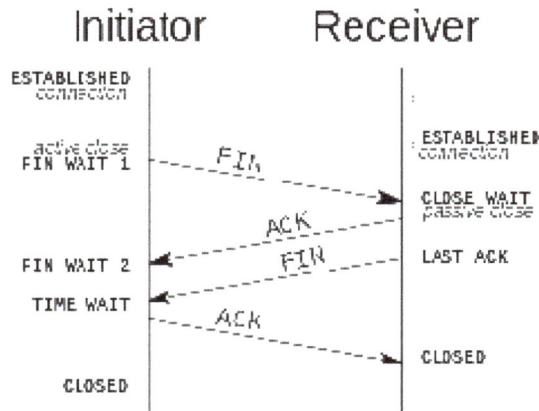

Figure 5. TCP connection

3.1.2. User Datagram Protocol (UDP)

User Datagram Protocol (UDP) is connectionless, so handshaking is not used in this service before two process start to communicate as shown in Figure 6. Also, UDP provides unreliable and unordered data transfer. There is no guarantee the messages sending to any loss. It has small segment header. Also, UDP doesn't have congestion control mechanism, so it is very fast, so useful for some applications.

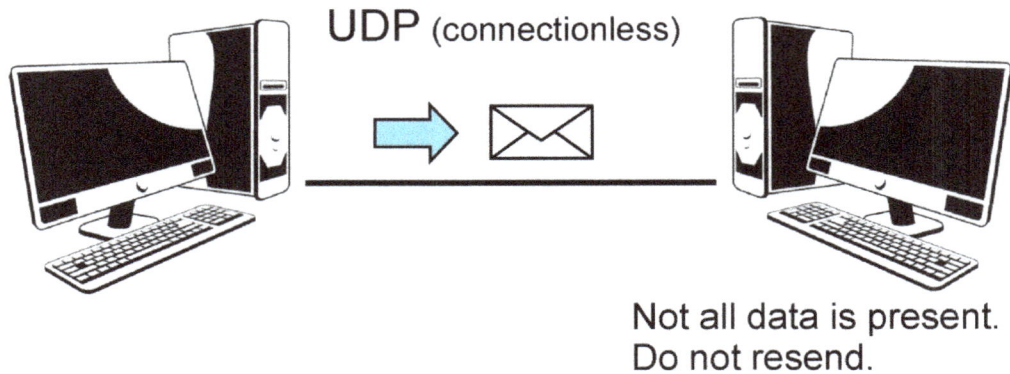

Figure 6. UDP (connectionless)

UDP and TCP service models have many different features as shown in Table 1.

Table 1. TCP and UDP features

TCP	UDP
Sequenced	Unsequenced
Reliable	Unreliable
Connection-oriented	Connectionless
Virtual circuit	Low overhead
Acknowledgments	No acknowledgment
Windowing flow control	No windowing or flow control

4. METHODS FOR FILE TRANSFER

4.1. The File Transfer Protocol (FTP)

The File Transfer Protocol was appeared by Abhay Bhushan and brought out as RFC 114 on 16 April 1971.It was changed by a TCP/IP version in June 1980 as RFC 765, and changed in October 1985 as RFC 959. RFC 959 is the current version and it's expanded by the some standards which are RFC 2228, RFC 2428. RFC 2228 is for security expansion and RFC 2428 is for providing support for IPv6 and establishes new passive mode.

The File Transfer Protocol is a standard network protocol which allows transferring computer files from one host to another host through a TCP based network [4]. It was came out to allow allocation of files and support to enhance remote computer's usage. FTP based on client server architecture and manages various control and data connections among client and the server as shown in Figure 7. Clear text sign in protocols are used by clients, which allows user to have password and username also connection without them (anonymously). SSL, TLS (FTPS) and SFTP secure the FTP.

Commands of FTP;

USER username: Used to send the user identification to the server.
PASS password: Used to send the user password to the server.
LIST: Used to ask the server to send back a list of all the files in the current remote directory. The list of files is sent over a (new and nonpersistent) data connection rather than the control TCP connection.
RETR filename: Used to retrieve (that is, get) a file from the current directory of the remote host. Triggers the remote host to initiate a data connection and to send the request files over the data connection.

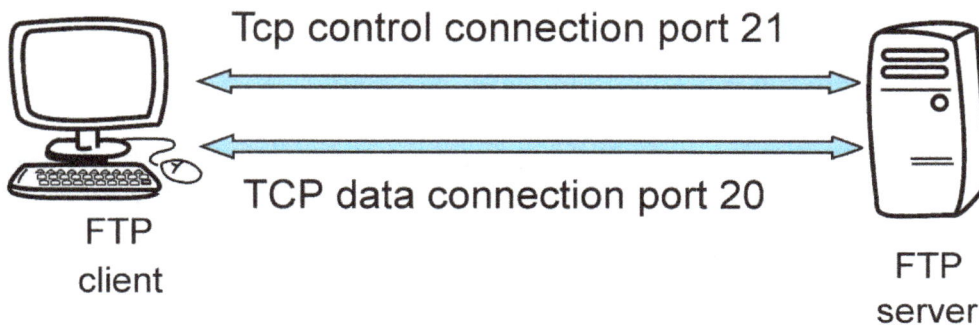

Figure 7. Control and data connection

FTP replies;

331 Username OK, password required

125 Data connection already open; transfer starting

425 can't open data connection

452 Error writing file

4.1.1. FTP Communication and Data Transfer

There are two modes for constituting data connection such as active and passive, FTP uses both modes. TCP control connection from a non-specific port N to the FTP server command port 21 will be created by client as shown in Figure 8. In active modes, incoming data connections on port N+1 from the server will be received by client. Port N+1 is used for report the server when FTP command is sent by client. In passive mode there is client behind a firewall and also not available for incoming TCP connections. In this mode PASV command is sent to the server by client through the control connection. Afterwards server sends a server IP address and server port number. In this way client can use it for opening a data connection from random client port to the server IP address and server port number [5]. Active mode and passive mode were updated to support IPv6 in 1998.

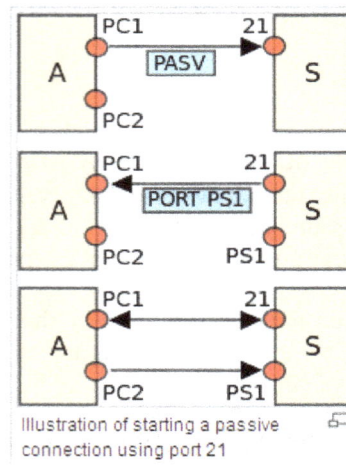

Figure 8. Illustration of starting a passive connection using port 21

The server reply with three digit status codes in ASCII upon control connection. These messages consist 100 Series, 200 Series, 300 Series, 400 Series, 500 Series etc. For example 200 OK means last command was successful. The numbers describe the code for the response and optional text describes explanation for humans. The current transfer may interrupt while data transferring. There are 4 representations can be used:

ASCII Mode: Data is converted. From sender's character representation to 8 bit ASCII before the transmission and to the receiver's character representation if there is any need. If files have data differently from plain text it will be unsuitable.

Image Mode: (Binary) Sender sends file byte for byte and receiver stores byte stream.

EBCDIC Mode: It uses EBCDIC characters among hosts for plain text.

Local Mode: It allows two computers to send data without convert them ASCII in a proprietary format.

For text files, there are various format control and record structure options. For the files which are containing Telnet or ASA, there are features to promote them.

Stream Mode : In this mode data is sent as persistent stream which is reliving FTP and making FTP trim from any processing.

Block Mode : In this mode data will break into different blocks. And will be passed on to TCP.
Compressed Mode : In this mode data is compacted using a single algorithm.

4.1.2 Basic Operations of FTP

FTP is founded on a client-server architecture which clients are transferring files to a server and receiving files from a server as shown in Figure 9. FTP period consists two connections which are transmitting standard FTP commands, responses and transferring the actual data.

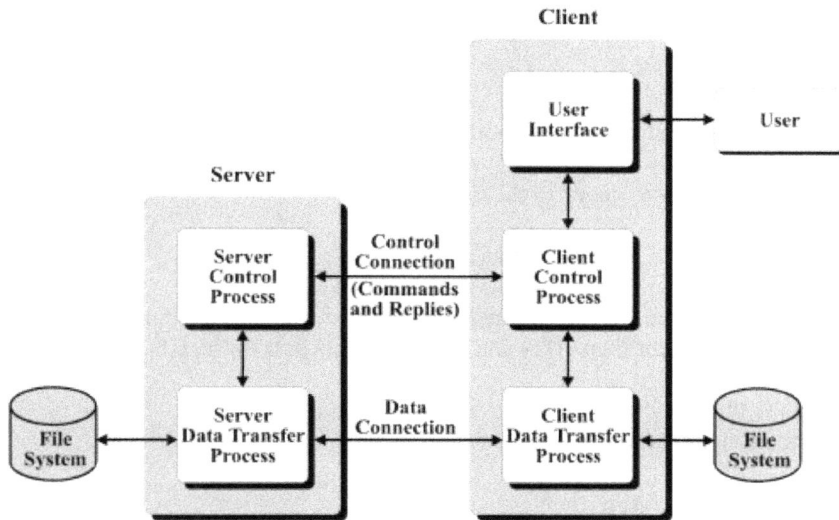

Figure 9. FTP Client-Server Application

The Client Control Process starts the Control Connection and in this process FTP commands and responses are transmit as shown in Figure 10. So that data connection will begin and will be used as needed. The Control Connection should be stay open mode during the data transferring time [6]. If any collision occurs during the data transfer FTP Data connection will be closed and session will be failed. There are parameters such as transfer mode, file structure, data representations during the data transfer and they are sent by FTP commands. When operation and parameters are transmitted, client will look up predefined TCP port and server.

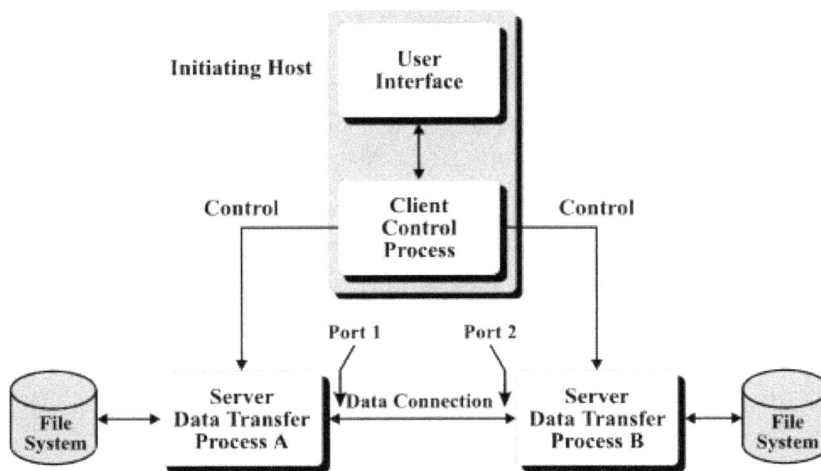

Figure 10. FTP Client Server Process

There is no Data connection need for between server and host in FTP. A FTP user process must guarantee that one of the third party machines starts on a designated port and other starts the Data Connection. In this module user creates a control connection and organize for the founding the Data connection among servers. At the figure below explains this scenario.

There is an alternative way which allows to users to start data connections and which also allows to greater network administrator. This is called FTP Passive Open.

4.1.2.1 Active FTP step by step

Client connect to FTP server with command port number 21.

FTP sends message and username query to the user

> Client enters connection information as needed.
> Server controls the information of sender and replies user.
> If the information is true, FTP command line will showed to the client.
> Client creates port bigger than 1024 and announce this port to the FTP server.

FTP server connects with this port number and transferring starts.
Client sends confirmation message.

4.1.2.2 Passive FTP step by step

Client connects to FTP server with command port number 21.

FTP server sends message and username query to the user.

> Client enters information of connection as needed.
> Server controls the information of sender and replies user.
> If the information is true, FTP client wait for the additional port from the server through PSAV command.

FTP client connects this port and starts to data transfer. Client sends a confirmation message.

4.1.2.3 Active FTP and Firewall

Even FTP user creates a port on its side and if there is a firewall, there would be a problem because that firewall will not give permission for this port as shown in Figure 11.

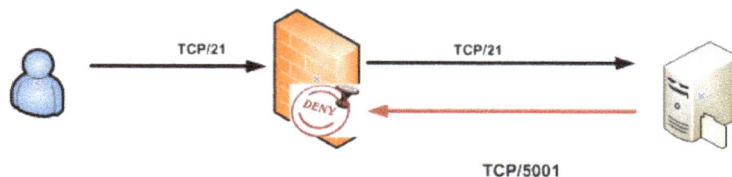

Figure 11. Active FTP and Firewall

4.1.2.4 Passive FTP and Firewall

In this module, even if FTP server opens additional port and if there is firewall, permission is needed from the firewall to access this port as shown in Figure 12.

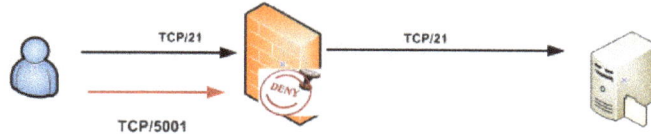

Figure 12. Passive FTP and Firewall

4.2. Secure File Transfer Protocol (SFTP)

There are methods for securing FTP such as FTPS and SFTP. Explicit FTPS is a FTP standard which allows users to demand encrypted FTP session. This session can be done with AUTH TLS command. Connections which are requesting TLS can be allow or deny because of server has these both options. There is a recommended standard called RFC 4217. Implicit FTPS is unconfirmed standard for FTP that needed the usage of SSL or TLS connection. It is defined for using different ports than plain FTP. Secure File transfer protocol which is also called as secure FTP is a protocol that transmits files and set of command for users, but its rest on different software technology as shown in Figure 13. SFTP uses the Secure Shell protocol to transmit files. It encrypts commands and data in terms of avoiding passwords and information which are susceptible. It cannot run with FTP software.

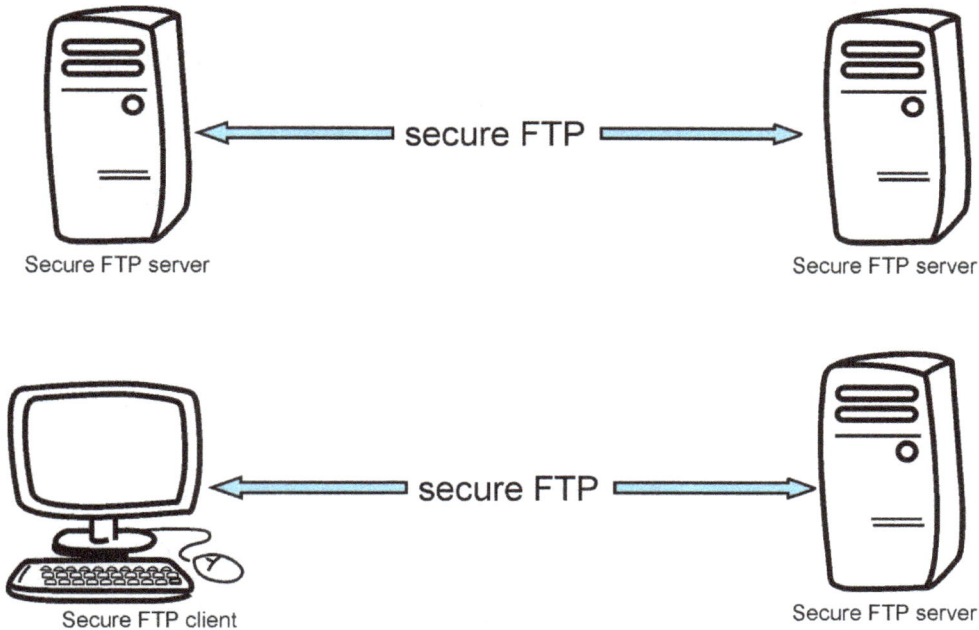

Figure 13. Secure FTP

5. METHODOLOGY

Before start working on this project, first searching programs like zip, gzip pack big data with size of scale is an issue. Inputs are analysed and various tests are run on data. Speed of data transfer are tried to control with Duplicati program which is preferred nowadays. Duplicati is an open source software because of that we will analyse codes and will learn what kind of algorithm is processing. Our aim is writing this program with C language which is the closest language of machine. Because we want to benefit from machine's hardware when we need it. We think that C language is most efficient language when it is used with efficient way, our gain will be a lot.

5.1. Scenario

For example, we want to send the file through FTP Server, we need to select FTP Button, and then we should continue with entering hostname, username and password. Then select the file we want to transfer. After that choose the backup path destination from the button of backup path. Click the send button. Our timer will start to work .Algorithm which is running at the background, controls the format of the file if it's compressed file or not. For example .mp3 and .mp4 is the compressed file format. If it's not compressed format, algorithm make the compression of the file. If it's already in the compressed format, then format will stay at the same format. After the process is completed, we are starting to send the file. Sending algorithm take two streams, one from server side, the other one is from file side. Then pieces of 1024 bytes of data are written on the server using streams. After the completing the sending file if it's a successful process, timer, streams and connection from server are closed. You can display the file on the server.

For example, we want to send the file through SFTP Server, we need to select SFTP Button, and then we should enter hostname, username and password. File is needed to select by user which we want to send. Using the backup path destination button, we will choose the destination of compressed file. Click the send button. Our timer will start to running. Algorithm which is running at the background controls the format of the file if it's compressed file or not. For example .mp3 and .mp4 is the compressed file format. If its not compressed format, algorithm make the compression of the file. If it's already in the compressed format, then format will not change. After the process is completed, sending process will begin. Sending algorithm take two streams, one from the server side and the other one is from file side. Then 1024 of bytes of data pieces is written on the server with usage of two streams. After the finishing the sending file, if it's a successful process, timer, streams and connection from server will closed. You can display the file on the server.

This Figure 14 is our project use cases diagram. It shows how application works. First of all, our main menu includes two file transfer protocols which are FTP and SFTP. User 1 chooses protocol one of them. Then, He or she continues to enter hostname, username and password such as hostname: 78.185.128.224, username: Ahmet FTP and password: 10. Moreover, the user chooses the sending file and send it. If, there is not any problem, user 1 can send the file successfully and other user 2 takes the sending file. However, if there is any problem such as loss connection, power supply problem or etc., he or she has to go back to enter the hostname, username and password menu and continue to send the file again.

Project main menu is shown in Figure 15. We have two choices which are File Transfer Protocol (FTP) and Secure File Transfer Protocol (SFTP). Then, we choose one of them to send the file and click the "Next" button. Otherwise, we click the "Exit" button and close the application as shown in Figure 15.

Selection of FTP or SFTP file transfer is shown in Figure 15, If the user wants to send file to FTP Server, the user should use FTP button. If the user wants to send file to SFTP Server, the user should use SFTP button.

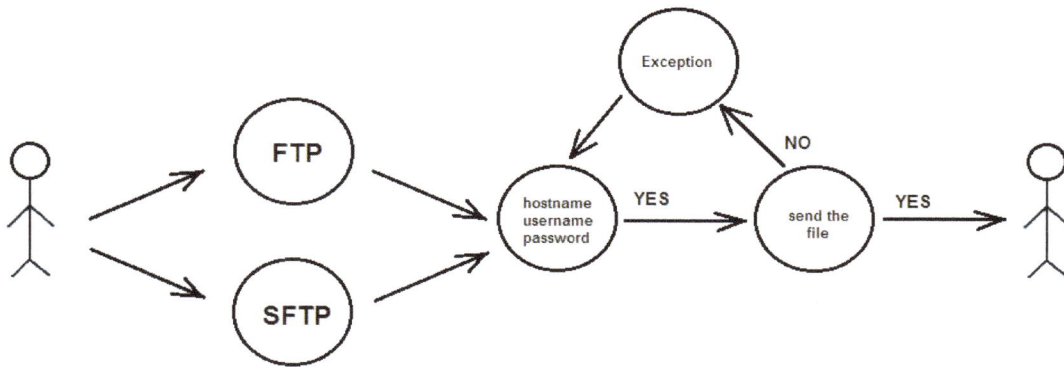

Figure 14. Use case Diagram for Huge Data File Transfer.

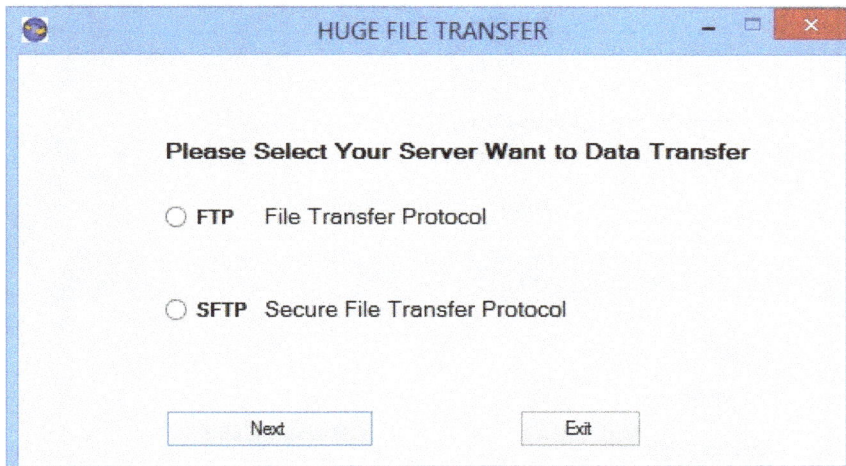

Figure 15. Select of SFTP or FTP

In Figure 16, our project flow chart diagram shows our application working steps. When, we start the application, we select FTP or SFTP. Next, we enter host, username and password. Then, if there is not a problem sending the file is successful. Otherwise, we enter the host, username and password again and continue to send file.

As shown in Figure 17, FTP authorization can be established by entering the necessary information.

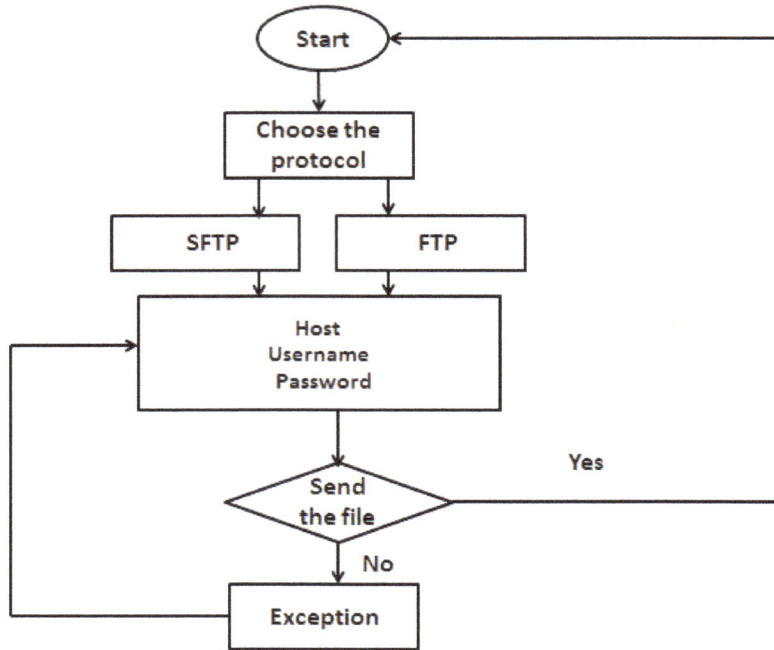

Figure 16. Flow Chart

Figure 17. FTP authorization

This is FTP menu includes host, username, password. User enters the host, username and password. Then, he or she clicks The "SELECT" button and select which file will send. Also, user clicks the "BACKUP PATH" button to put the copy of this file zip format. If he or she wants to go back, user clicks the "BACK" button. In addition, if user wants to stop the sending file, he or she clicks "STOP" button. Then, user, send the file to click "SEND" button as shown in Figure 17. Finally, user can see how much time file sends.

Hostname: The IP of FTP Server
Username: Name of user for login to server.
Password: Password of user account.
Select button: The File which we want to send.

Backup Path Button: Path of compressed file's destination.
Send Button: Button for sending file to the server.
Stop Button: Button for stoping file transfer.
Back Button: Turn the selection page of choosing FTP or SFTP.

When user chooses the FTP selection, we use 21 port and program converts the file zip format. However, if file is zip, mp3, mp4 format, program doesn't convert this file because these are compressed file. Then, file divides 1024 bytes packets and sends it and other user takes the file zip format.

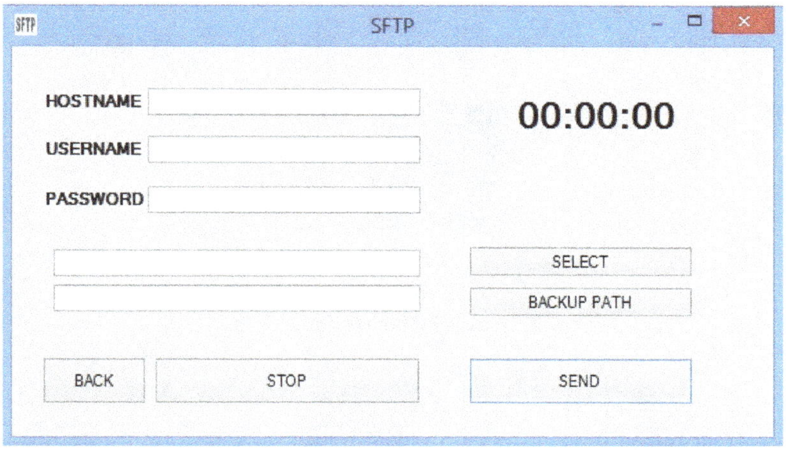

Figure 18. SFTP authorization

This is another selection to send the file which is SFTP as shown in figure 18. User enters hostname, username and password. The "SELECT" button and select which file will send. Also, user clicks the "BACKUP PATH" button to put the copy of this file zip format. If he or she wants to go back, user clicks the "BACK" button. In addition, if user wants to stop the sending file, he or she clicks "STOP" button. Finally, user, send the file to click "SEND" button. Finally, user can see how much time file sends.

If user selects the SFTP protocol, we use 22 ports. Firstly, file is converted the zip format exception of the zip, mp3 and mp4 format files. In addition, user sends the file and another user takes the file. This selection sending time is longer than FTP. Because, this is securely.

Hostname: The IP of SFTP Server
Username: Name of user for login to server.
Password: Password of user account.
Select button: The File which we want to send.
Backup Path Button: Path of compressed file's destination.
Send Button: Button for sending file to the server.
Stop Button: Button for stoping file transfer.
Back Button: Turn the selection page of choosing FTP or SFTP.

6. PERFORMANCE COMPARISON OF PROPOSED SYSTEM AND DUPLICATI

In Figure 19, we showed the program time speed between Duplicati for 100 Mb file in FTP protocol. Our program is faster than Duplicati.

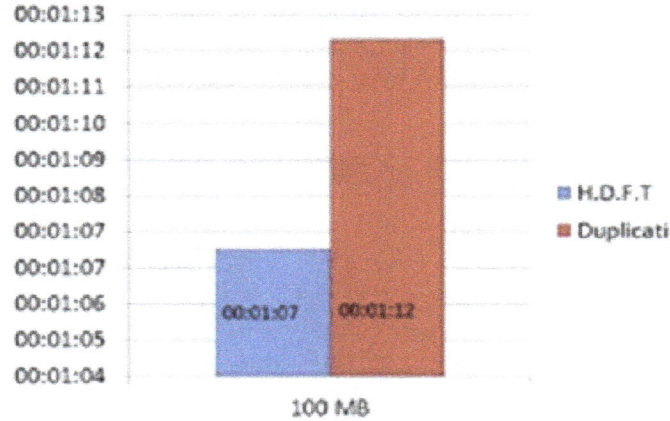

Figure 19. 100 Mb for FTP

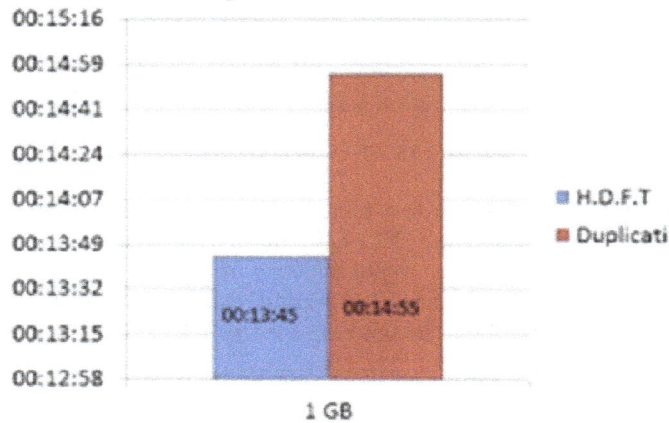

Figure 20. 1GB for FTP

In Figure 20, we tested the file sending to use FTP for 1 GB between our program and Duplicati. Our program sends the file shorter time than Duplicati.

We compared the sending file time our program to use FTP with Duplicati for 10 GB as shown in Figure 21. Our program sends the file faster than other.

We sent the 100 mb file both our program and Duplicati in SFTP protocol as shown in Figure 22. Our program sends the file very short time.

We sent the 1 GB file between Duplicati and our program in SFTP as shown in Figure 23. We sent the file faster than Duplicati.

Finally, we sent the 10 GB file to use our program and Duplicati in with SFTP protocol as shown in Figure 24. Our program takes very short time.

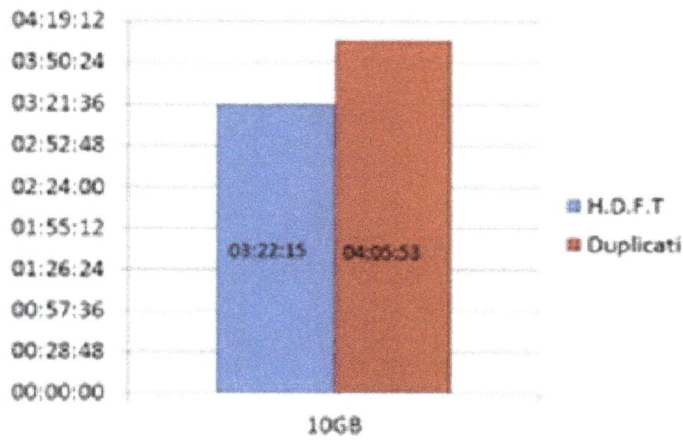

Figure 21. 10 GB for FTP

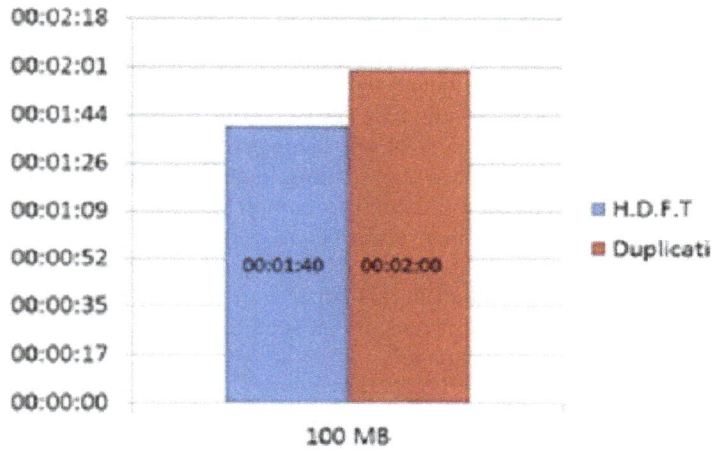

Figure 22. 100 Mb for SFTP

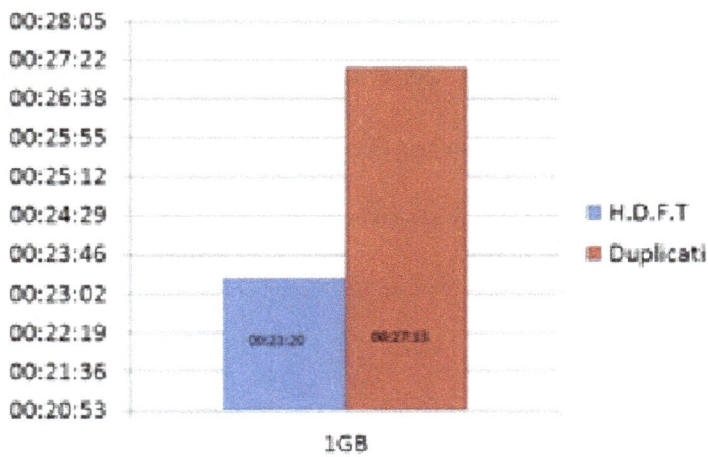

Figure 23. 1 GB for SFTP

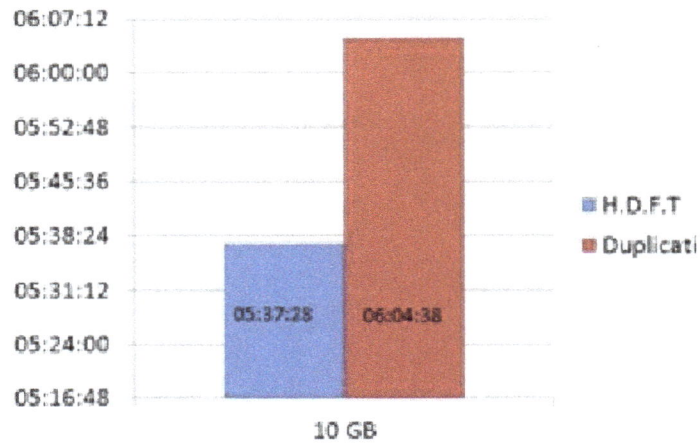

Figure 24. 10 GB for SFTP

3. CONCLUSIONS

The first aim of this paper is to provide a better and more efficient performance and to develop a faster application than Duplicati. First of all, the algorithm of Duplicati is examined. Then, we develop and create our own algorithms for sending huge data file. After obtaining the test result, we assure huge data file transfer application to complete the same process faster than Duplicati less than two times. So, huge data file transfer becomes more effective and faster.

The algorithm, which is implemented in this paper, consists compressing the file with the format of .zip and splitting the file into the pieces. It can send the file successfully, efficiently and faster. This is the main contribution of this paper.

REFERENCES

[1]   Information Storage and Management Storing, Managing, and Protecting Digital Information Edited by G. Somasundaram Alok Shrivastava, EMC Education Services, 27-29.

[2]   The Embedded Internet TCP/IP basics, implementation and applications Edited by Sergio Scaglia , 2007,225-228

[3]   Computer Networking A Top-Down Approach Featuring the Internet third edition Edited by James F. Kurose, Keith W. Ross 96-98.

[4]   Data Transfer   Linda Woodard Consultant Cornell CAC Workshop: Parallel Computing on Stampede: June 18, 2013, 2.

[5]   Managed File Transfer Solutions using DataPower and WebSphere MQ File Transfer Edition Edited by IBM,4-5.

[6]   Computing Community Consortium Version 8 : December 22 , 2008

13

NONLINEAR CHANNEL ESTIMATION FOR OFDM SYSTEM BY COMPLEX LS-SVM UNDER HIGH MOBILITY CONDITIONS

Anis Charrada[1] and Abdelaziz Samet[2]

[1 and 2] CSE Research Unit, Tunisia Polytechnic School, Carthage University, Tunis, Tunisia .
[1]anis.charrada@gmail.com, [2]abdelaziz.samet@ept.rnu.tn

ABSTRACT

A nonlinear channel estimator using complex Least Square Support Vector Machines (LS-SVM) is proposed for pilot-aided OFDM system and applied to Long Term Evolution (LTE) downlink under high mobility conditions. The estimation algorithm makes use of the reference signals to estimate the total frequency response of the highly selective multipath channel in the presence of non-Gaussian impulse noise interfering with pilot signals. Thus, the algorithm maps trained data into a high dimensional feature space and uses the structural risk minimization (SRM) principle to carry out the regression estimation for the frequency response function of the highly selective channel. The simulations show the effectiveness of the proposed method which has good performance and high precision to track the variations of the fading channels compared to the conventional LS method and it is robust at high speed mobility.

Keywords

Complex LS-SVM, Mercer's kernel, nonlinear channel estimation, impulse noise, OFDM, LTE.

1. INTRODUCTION

Channel estimation in wireless OFDM systems is an active research area, especially in the case of frequency selective time varying multipath fading channels. Several estimation algorithms have been developed, such that LS [1], MMSE [2] and estimation with decision feedback [3]. Channel estimation by neural network is also described in [4]. However, in a practical environment where non-Gaussian impulse noise can be present, the classical estimation methods may not be effective for this impulse noise.

The use of Support Vector Machines (SVMs) has already been proposed to solve a variety of signal processing and digital communications problems. Signal equalization and detection for multicarrier MC-CDMA system is presented in [5]. Also, adaptive multiuser detector for direct sequence CDMA signals in multipath channels is developed in [6]. In all these applications, SVM methods outperform classical approaches due to its improved generalization capabilities.

Here, a proposed SVM robust version for nonlinear channel estimation in the presence of non-Gaussian impulse noise that is specifically adapted to pilot-aided OFDM structure is presented. In fact, impulses of short duration are unpredictable and contain spectral components on all subchannels which impact the decision of the transmitted symbols on all subcarriers.

The channel estimation algorithm is based on the nonlinear least square support vector machines (LS-SVM) method in order to improve communication efficiency and quality of OFDM systems. The principle of the proposed nonlinear LS-SVM algorithm is to exploit the information provided by the reference signal to estimate the channel frequency response. In highly selective multipath fading channel, where complicated nonlinearities can be present, the estimation precision can be lowed by using linear method. So, we adapt the nonlinear LS-SVM algorithm which transforms the nonlinear estimation in low dimensional space into the linear estimation in high dimensional space, so it improves the estimation precision.

In this contribution, the proposed nonlinear complex LS-SVM technique is applied to LTE downlink highly selective channel using pilot symbols. For the purpose of comparison with conventional LS algorithm, we develop the nonlinear LS-SVM algorithm in terms of the RBF kernel. Simulation section illustrates the advantage of this algorithm over LS algorithm in high mobility environment. The nonlinear complex LS-SVM method shows good results under high mobility conditions due to its improved generalization ability.

The scheme of the paper is as follows. Section 2 briefly introduces the OFDM system model. We present the formulation of the proposed nonlinear complex LS-SVM channel estimation method in section 3. Section 4 presents the simulation results when comparing with LS standard algorithm. Finally, in section 5, conclusions are drawn.

2. SYSTEM MODEL

The OFDM system model consists firstly of mapping binary data streams into complex symbols by means of QAM modulation. Then data are transmitted in frames by means of serial-to-parallel conversion. Some pilot symbols are inserted into each data frame which is modulated to subcarriers through IDFT. These pilot symbols are inserted for channel estimation purposes. The IDFT is used to transform the data sequence $X(k)$ into time domain signal as follow:

$$x(n) = IDFT_N\{X(k)\} = \sum_{k=0}^{N-1} X(k)\, e^{j\frac{2\pi}{N}kn}, \qquad n = 0, \cdots, N-1 \qquad (1)$$

One guard interval is inserted between every two OFDM symbols in order to eliminate inter-symbol interference (ISI). This guard time includes the cyclically extended part of the OFDM symbol in order to preserve orthogonality and eliminate inter-carrier interference (ICI). It is well known that if the channel impulse response has a maximum of L resolvable paths, then the GI must be at least equal to L [7].

Thus, for the OFDM system comprising N subcarriers which occupy a bandwidth B, each OFDM symbol is transmitted in time T and includes a cyclic prefix of duration T_{cp}. Therefore, the duration of each OFDM symbol is $T_u = T - T_{cp}$. Every two adjacent subcarriers are spaced by $\delta f = 1/T_u$. The output signal of the OFDM system is converted into serial signal by parallel to serial converter. A complex white Gaussian noise process $N\,(0, \sigma_w^2)$ with power spectral density $N_0/2$ is added through a frequency selective time varying multipath fading channel.

In a practical environment, impulse noise can be present, and then the channel becomes nonlinear with non Gaussian impulse noise. The impulse noise can significantly influence the performance of the OFDM communication system for many reasons. First, the time of the arrival of an impulse is unpredictable and shapes of the impulses are not known and they vary considerably. Moreover, impulses usually have very high amplitude, and thus high energy, which can be much greater than the energy of the useful signal [8].

The impulse noise is modeled as a Bernoulli-Gaussian process and it was generated with the Bernoulli-Gaussian process function $i(n) = v(n)\lambda(n)$ where $v(n)$ is a random process with Gaussian distribution and power σ_{BG}^2, and where $\lambda(n)$ is a random process with probability [9]

$$P_r(\lambda(n)) = \begin{cases} p & \lambda = 1 \\ 1-p, & \lambda = 0. \end{cases} \qquad (2)$$

At the receiver, and after removing guard time, the discrete-time baseband OFDM signal for the system including impulse noise is

$$y(n) = \sum_{k=0}^{N-1} X(k)H(k)e^{j\frac{2\pi}{N}kn} + w(n) + i(n), \qquad n = 0, \cdots, N-1 \qquad (3)$$

where $y(n)$ are time domain samples and $H(k) = DFT_N\{h(n)\}$ is the channel's frequency response at the kth frequency. The sum of both terms of the AWGN noise and impulse noise constitute the total noise given by $z(n) = w(n) + i(n)$.

Let Ω_P the subset of N_P pilot subcarriers and ΔP the pilot interval in frequency domain. Over this subset, channel's frequency response can be estimated, and then interpolated over other subcarriers $(N - N_P)$. These remaining subchannels are interpolated by the nonlinear complex LS-SVM algorithm. The OFDM system can be expressed as

$$y(n) = y^P(n) + y^D(n) + z(n)$$

$$= \sum_{k\in\{\Omega_P\}} X^P(k)H(k)e^{j\frac{2\pi}{N}kn} + \sum_{k\notin\{\Omega_P\}} X^D(k)H(k)e^{j\frac{2\pi}{N}kn} + z(n) \qquad (4)$$

where $X^P(k)$ and $X^D(k)$ are complex pilot and data symbol respectively, transmitted at the kth subcarrier. Note that, pilot insertion in the subcarriers of every OFDM symbol must satisfy the demand of the sampling theory and uniform distribution [10].

After DFT transformation, $y(n)$ becomes

$$Y(k) = DFT_N\{y(n)\} = \frac{1}{N}\sum_{n=0}^{N-1} y(n)\, e^{-j\frac{2\pi}{N}kn}, \quad k = 0, \cdots, N-1 \qquad (5)$$

Assuming that ISI are eliminated, therefore

$$Y(k) = X(k)H(k) + W(k) + I(k) = X(k)H(k) + e(k), \quad k = 0, \cdots, N-1 \qquad (6)$$

where $e(k)$ represents the sum of the AWGN noise $W(k)$ and impulse noise $I(k)$ in the frequency domain, respectively.

(6) may be presented in matrix notation

$$Y = XFh + W + I = XH + e \qquad (7)$$

where

$$\begin{aligned}
X &= diag\big(X(0), X(1), \cdots, X(N-1)\big) \\
Y &= [Y(0), \cdots, Y(N-1)]^T \\
W &= [W(0), \cdots, W(N-1)]^T \\
I &= [I(0), \cdots, I(N-1)]^T \\
H &= [H(0), \cdots, H(N-1)]^T \\
e &= [e(0), \cdots, e(N-1)]^T
\end{aligned}$$

$$F = \begin{bmatrix} W_N^{00} & \cdots & W_N^{0(N-1)} \\ \vdots & \ddots & \vdots \\ W_N^{(N-1)0} & \cdots & W_N^{(N-1)(N-1)} \end{bmatrix}$$

and

$$W_N^{i,k} = \left(\frac{1}{\sqrt{N}}\right) exp^{-j2\pi\left(\frac{ik}{N}\right)}. \qquad (8)$$

3. NONLINEAR COMPLEX LS-SVM ESTIMATOR

First, let the OFDM frame contains Ns OFDM symbols which every symbol includes N subcarriers. As every OFDM symbol has N_P uniformly distributed pilot symbols, the transmitting pilot symbols are $X^P = diag(X(s, m\,\Delta P)), m = 0, 1, \cdots N_P - 1$, where s and m are labels in time domain and frequency domain respectively.

The proposed channel estimation method is based on nonlinear complex LS-SVM algorithm which has two separate phases: training phase and estimation phase. In training phase, we estimate first the subchannels pilot symbols according to LS criterion to strike $min\ [(Y^P - X^P Fh)(Y^P - X^P Fh)^H]$ [11], as

$$\hat{H}^P = X^{P^{-1}} Y^P \qquad (9)$$

where $Y^P = Y(s, m\,\Delta P)$ and $\hat{H}^P = \hat{H}(s, m\,\Delta P)$ are the received pilot symbols and the estimated frequency responses for the sth OFDM symbol at pilot positions $m\,\Delta P$, respectively. Then, in the estimation phase and by the interpolation mechanism, frequency responses of data subchannels can be determined. Therefore, frequency responses of all the OFDM subcarriers are

$$\hat{H}(s, q) = f\left(\hat{H}^P(s, m\,\Delta P)\right) \qquad (10)$$

where $q = 0, \cdots, N - 1$, and $f(\cdot)$ is the interpolating function, which is determined by the nonlinear complex LS-SVM approach.

In high mobility environments, where the fading channels present very complicated nonlinearities especially in deep fading case, the linear approaches cannot achieve high estimation precision. Therefore, we adapt here a nonlinear complex LS-SVM technique since SVM is superior in solving nonlinear, small samples and high dimensional pattern recognition [10]. Therefore, we map the input vectors to a higher dimensional feature space $\mathcal{H}$ (possibly infinity) by means of nonlinear transformation $\boldsymbol{\varphi}$. Thus, the regularization term is referred to the regression vector in the RKHS. The following regression function is then

$$\hat{H}(m\,\Delta P) = \boldsymbol{w}^T \boldsymbol{\varphi}(m\,\Delta P) + b + e_m, \qquad m = 0, \cdots N_P - 1 \qquad (11)$$

where $\boldsymbol{w}$ is the weight vector, b is the bias term well known in the SVM literature and residuals $\{e_m\}$ account for the effect of both approximation errors and noise. In the SVM framework, the optimality criterion is a regularized and constrained version of the regularized LS criterion. In general, SVM algorithms minimize a regularized cost function of the residuals, usually the Vapnik's $\varepsilon - insensitivity$ cost function [9]. A robust cost function is introduced to improve the performance of the estimation algorithm which is ε -Huber robust cost function, given by [12]

$$\mathcal{L}^\varepsilon(e_m) = \begin{cases} 0, & |e_m| \leq \varepsilon \\ \dfrac{1}{2\gamma}(|e_m| - \varepsilon)^2, & \varepsilon \leq |e_m| \leq e_C \\ C(|e_m| - \varepsilon) - \dfrac{1}{2}\gamma C^2, & e_C \leq |e_m| \end{cases} \qquad (12)$$

where $e_C = \varepsilon + \gamma C$, ε is the insensitive parameter which is positive scalar that represents the insensitivity to a low noise level, parameters γ and C control essentially the trade-off between the regularization and the losses, and represent the relevance of the residuals that are in the linear or in the quadratic cost zone, respectively. The cost function is linear for errors above e_C, and quadratic for errors between ε and e_C. Note that, errors lower than ε are ignored in the

$\varepsilon - insensitivite$ zone. On the other hand, the quadratic cost zone uses the $L_2 - norm$ of errors, which is appropriate for Gaussian noise, and the linear cost zone limits the effect of sub-Gaussian noise [13]. Therefore, the ε -Huber robust cost function can be adapted to different kinds of noise.

Since e_m is complex, let $\mathcal{L}^\varepsilon(e_m) = \mathcal{L}^\varepsilon(\mathcal{R}(e_m)) + \mathcal{L}^\varepsilon(\mathfrak{I}(e_m))$, where $\mathcal{R}(\cdot)$ and $\mathfrak{I}(\cdot)$ represent real and imaginary parts, respectively.

Now, we can state the primal problem as minimizing

$$\frac{1}{2}\|w\|^2 + \frac{1}{2\gamma}\sum_{m\in I_1}(\xi_m + \xi_m^*)^2 + C\sum_{m\in I_2}(\xi_m + \xi_m^*) + \frac{1}{2\gamma}\sum_{m\in I_3}(\zeta_m + \zeta_m^*)^2$$
$$+ C\sum_{m\in I_4}(\zeta_m + \zeta_m^*) - \frac{1}{2}\sum_{m\in I_2, I_4}\gamma C^2 \tag{13}$$

constrained to

$$\mathcal{R}\big(\widehat{H}(m\,\Delta P) - w^T \varphi(m\,\Delta P) - b\big) \leq \varepsilon + \xi_m$$

$$\mathfrak{I}\big(\widehat{H}(m\,\Delta P) - w^T \varphi(m\,\Delta P) - b\big) \leq \varepsilon + \zeta_m$$

$$\mathcal{R}\big(-\widehat{H}(m\,\Delta P) + w^T \varphi(m\,\Delta P) + b\big) \leq \varepsilon + \xi_m^*$$

$$\mathfrak{I}\big(-\widehat{H}(m\,\Delta P) + w^T \varphi(m\,\Delta P) + b\big) \leq \varepsilon + \zeta_m^*$$

$$\xi_m^{(*)}, \zeta_m^{(*)} \geq 0 \tag{14}$$

for $m = 0, \cdots, N_P - 1$, where ξ_m and ξ_m^* are slack variables which stand for positive and negative errors in the real part, respectively. ζ_m and ζ_m^* are the errors for the imaginary parts. I_1, I_2, I_3 and I_4 are the set of samples for which:

I_1 :   real part of the residuals are in the quadratic zone;

I_2 : : real part of the residuals are in the linear zone;

I_3 : : imaginary part of the residuals are in the quadratic zone;

I_4 : : imaginary part of the residuals are in the linear zone.

To transform the minimization of the primal functional (13) subject to constraints in (14), into the optimization of the dual functional, we must first introduce the constraints into the primal functional. Thus, the primal dual functional is as follow:

$$L_{Pd} = \frac{1}{2}\|w\|^2 + \frac{1}{2\gamma}\sum_{m\in I_1}(\xi_m + \xi_m^*)^2 + C\sum_{m\in I_2}(\xi_m + \xi_m^*) + \frac{1}{2\gamma}\sum_{m\in I_3}(\zeta_m + \zeta_m^*)^2$$
$$+ C\sum_{m\in I_4}(\zeta_m + \zeta_m^*) - \frac{1}{2}\sum_{m\in I_2, I_4}\gamma C^2 - \sum_{m=0}^{N_P-1}(\beta_m\xi_m + \beta_m^*\xi_m^*) - \sum_{m=0}^{N_P-1}(\lambda_m\zeta_m + \lambda_m^*\zeta_m^*)$$
$$+ \sum_{m=0}^{N_P-1}\alpha_{\mathcal{R},m}\big[\mathcal{R}\big(\widehat{H}(m\,\Delta P) - w^T\varphi(m\,\Delta P) - b\big) - \varepsilon - \xi_m\big]$$

$$+ \sum_{m=0}^{N_P-1} \alpha_{I,m} \left[\Im\big(\hat{H}(m\,\Delta P) - \boldsymbol{w}^T \boldsymbol{\varphi}(m\,\Delta P) - b\big) - j\varepsilon - j\zeta_m \right]$$

$$+ \sum_{m=0}^{N_P-1} \alpha_{\mathcal{R},m}^* \left[\mathcal{R}\big(-\hat{H}(m\,\Delta P) + \boldsymbol{w}^T \boldsymbol{\varphi}(m\,\Delta P) + b\big) - \varepsilon - \xi_m^* \right]$$

$$+ \sum_{m=0}^{N_P-1} \alpha_{I,m}^* \left[\Im\big(-\hat{H}(m\,\Delta P) + \boldsymbol{w}^T \boldsymbol{\varphi}(m\,\Delta P) + b\big) - j\varepsilon - j\zeta_m^* \right] \tag{15}$$

with the Lagrange multipliers (or dual variables) constrained to $\alpha_{\mathcal{R},m}, \alpha_{I,m}, \beta_m, \lambda_m, \alpha_{\mathcal{R},m}^*, \alpha_{I,m}^*,$ $\beta_m^*, \lambda_m^* \geq 0$ and $\xi_m, \zeta_m, \xi_m^*, \zeta_m^* \geq 0$.
According to Karush-Kuhn-Tucker (KKT) conditions [12]

$$\beta_m \xi_m = 0, \ \beta_m^* \xi_m^* = 0 \ \text{and} \ \lambda_m \zeta_m = 0, \lambda_m^* \zeta_m^* = 0. \tag{16}$$

Then, by making zero the primal-dual functional gradient with respect to ϖ_i, we obtain an optimal solution for the weights

$$\boldsymbol{w} = \sum_{m=0}^{N_P-1} \psi_m \, \boldsymbol{\varphi}(m\,\Delta P) = \sum_{m=0}^{N_P-1} \psi_m \, \boldsymbol{\varphi}(P_m) \tag{17}$$

where $\psi_m = (\alpha_{\mathcal{R},m} - \alpha_{\mathcal{R},m}^*) + j(\alpha_{I,m} - \alpha_{I,m}^*)$ with $\alpha_{\mathcal{R},m}, \alpha_{\mathcal{R},m}^*, \alpha_{I,m}, \alpha_{I,m}^*$ are the Lagrange multipliers for real and imaginary part of the residuals and $P_m = (m\,\Delta P), \ m = 0, \cdots, N_P - 1$ are the pilot positions.

We define the Gram matrix as

$$\boldsymbol{G}(u,v) = < \boldsymbol{\varphi}(P_u), \boldsymbol{\varphi}(P_v) > = K(P_u, P_v) \tag{18}$$

where $K(P_u, P_v)$ is a Mercer's kernel which represent in this contribution the RBF kernel matrix which allows obviating the explicit knowledge of the nonlinear mapping $\boldsymbol{\varphi}(\cdot)$. A compact form of the functional problem can be stated in matrix format by placing optimal solution $\boldsymbol{w}$ into the primal dual functional and grouping terms. Then, the dual problem consists of maximizing

$$-\frac{1}{2} \boldsymbol{\psi}^H (\boldsymbol{G} + \gamma \boldsymbol{I}) \boldsymbol{\psi} + \mathcal{R}\big(\boldsymbol{\psi}^H Y^P\big) - (\boldsymbol{\alpha}_{\mathcal{R}} + \boldsymbol{\alpha}_{\mathcal{R}}^* + \boldsymbol{\alpha}_I + \boldsymbol{\alpha}_I^*) \boldsymbol{1}\varepsilon \tag{19}$$

constrained to $0 \leq \alpha_{\mathcal{R},m}, \alpha_{\mathcal{R},m}^*, \alpha_{I,m}, \alpha_{I,m}^* \leq C$, where $\boldsymbol{\psi} = [\psi_0, \cdots, \psi_{N_P-1}]^T$; $\boldsymbol{I}$ and $\boldsymbol{1}$ are the identity matrix and the all-ones column vector, respectively; $\boldsymbol{\alpha}_{\mathcal{R}}$ is the vector which contains the corresponding dual variables, with the other subsets being similarly represented. The weight vector can be obtained by optimizing (19) with respect to $\alpha_{\mathcal{R},m}, \alpha_{\mathcal{R},m}^*, \alpha_{I,m}, \alpha_{I,m}^*$ and then substituting into (17).
Therefore, and after training phase, frequency responses at all subcarriers in each OFDM symbol can be obtained by SVM interpolation

$$\hat{H}(k) = \sum_{m=0}^{N_P-1} \psi_m \, K(P_m, k) + b \tag{20}$$

for $k = 1, \cdots, N$. Note that, the obtained subset of Lagrange multipliers which are nonzero will provide with a sparse solution. As usual in the SVM framework, the free parameter of the kernel and the free parameters of the cost function have to be fixed by some a priori knowledge of the problem, or by using some validation set of observations [9].

4. SIMULATION RESULTS

We consider the channel impulse response of the frequency-selective fading channel model which can be written as

$$h(\tau, t) = \sum_{l=0}^{L-1} h_l(t)\, \delta(t - \tau_l) \tag{21}$$

where $h_l(t)$ is the impulse response representing the complex attenuation of the lth path, τ_l is the random delay of the lth path and L is the number of multipath replicas. The specification parameters of an extended vehicular A model (EVA) for downlink LTE system with the excess tap delay and the relative power for each path of the channel are shown in table 1. These parameters are defined by 3GPP standard [14].

Table 1. Extended Vehicular A model (EVA) [14].

Excess tap delay [ns]	Relative power [dB]
0	0.0
30	-1.5
150	-1.4
310	-3.6
370	-0.6
710	-9.1
1090	-7.0
1730	-12.0
2510	-16.9

In order to demonstrate the effectiveness of our proposed technique and evaluate the performance, two objective criteria, the signal-to-noise ratio (SNR) and signal-to-impulse ratio (SIR) are used. The SNR and SIR are given by [9]

$$SNR_{dB} = 10log_{10}\left(\frac{E\{|y(n) - w(n) - i(n)|^2\}}{\sigma_w^2}\right) \tag{22}$$

and

$$SIR_{dB} = 10log_{10}\left(\frac{E\{|y(n) - w(n) - i(n)|^2\}}{\sigma_{BG}^2}\right) \tag{23}$$

Then, we simulate the OFDM downlink LTE system with parameters presented in table 2. The nonlinear complex LS-SVM estimate a number of OFDM symbols in the range of 140 symbols, corresponding to one radio frame LTE. Note that, the LTE radio frame duration is 10 ms [15], which is divided into 10 subframes. Each subframe is further divided into two slots, each of 0.5 ms duration.

For the purpose of evaluation the performance of the nonlinear complex LS-SVM algorithm under high mobility conditions, we consider a scenario for downlink LTE system for a mobile speed equal to 350 Km/h. Accordingly, we take into account the impulse noise with $p = 0.2$ which was added to the reference signals with different rates of SIR and it ranged from -10 to 10 dB.

Table 2. Parameters of simulations [15],[16] and [17].

Parameters	Specifications
OFDM system	LTE/Downlink
Constellation	16-QAM
Mobile Speed (Km/h)	350
T_s (µs)	72
f_c (GHz)	2.15
δf (KHz)	15
B (MHz)	5
Size of DFT/IDFT	512
Number of paths	9

Figure (1) presents the variations in time and in frequency of the channel frequency response for the considered scenario.

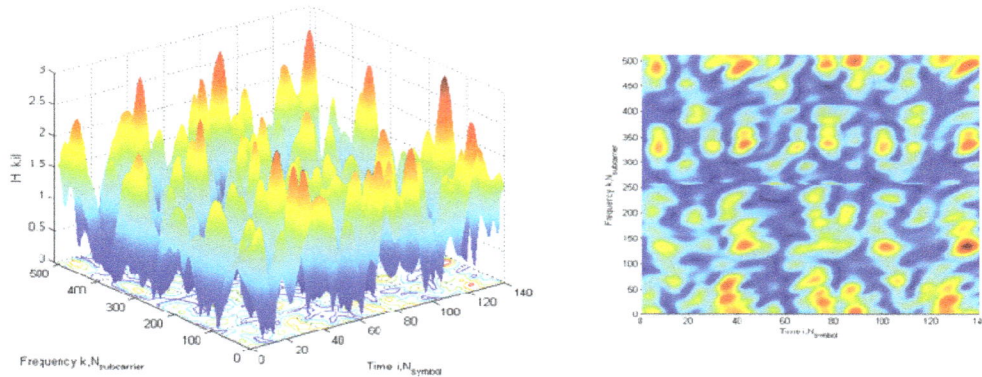

Figure 1. Variations in time and in frequency of the channel frequency response for mobile speed =350 Km/h.

Figure (2) shows the performance of the LS and nonlinear complex LS-SVM algorithms in the presence of additive Gaussian noise as a function of SNR for SIR=0 and -10 dB respectively. A poor performance is noticeably exhibited by LS and better performance is observed with nonlinear complex LS-SVM.

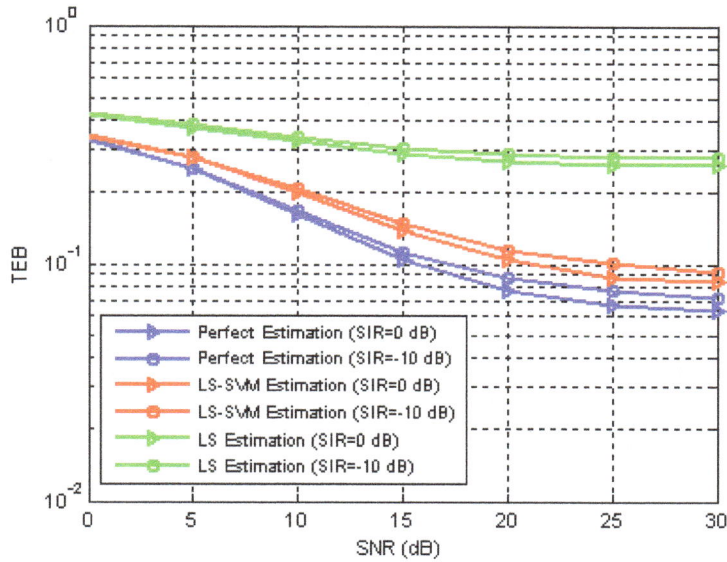

Figure 2. BER as a function of SNR for a mobile speed at 350 Km/h for SIR=0 and -10 dB with p=0.2.

Figure (3) presents a comparison between LS and nonlinear complex LS-SVM in the presence of additive impulse noise for SNR=20 dB. The comparison of these methods reveals that nonlinear complex LS-SVM outperform LS estimator in high mobility conditions especially for high SNR as confirmed by figure (4) for SNR=30 dB.

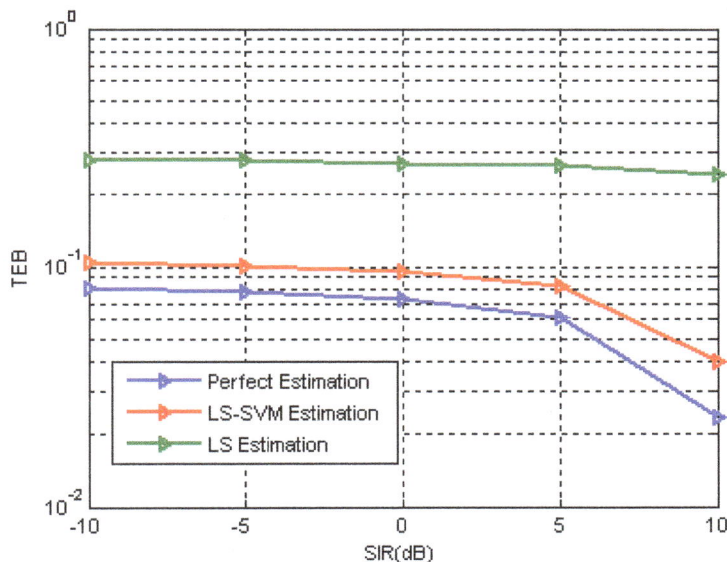

Figure 3. BER as a function of SIR for a mobile speed at 350 Km/h for SNR=20 dB with p=0.2.

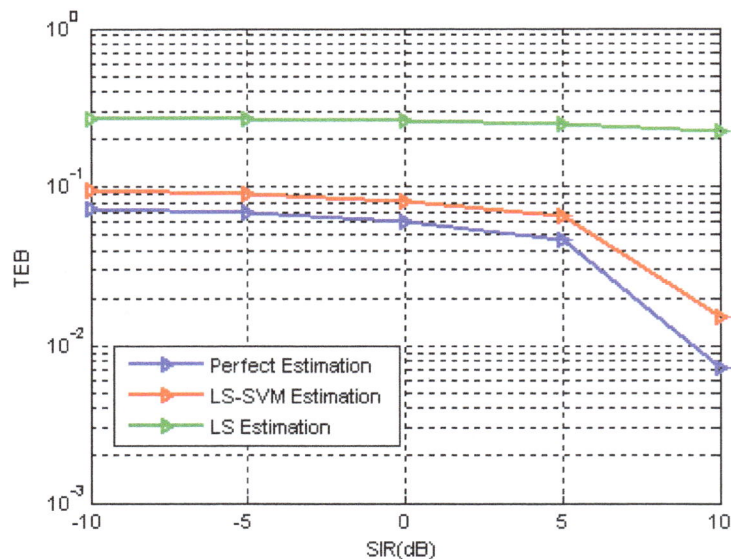

Figure 4. BER as a function of SIR for a mobile speed at 350 Km/h for SNR=30 dB with p=0.2.

5. CONCLUSION

In this contribution, we have presented a new nonlinear complex LS-SVM based channel estimation technique for a downlink LTE system under high mobility conditions in the presence of non-Gaussian impulse noise interfering with OFDM reference symbols.

The proposed method is based on learning process that uses training sequence to estimate the channel variations. Our formulation is based on nonlinear complex LS-SVM specifically developed for pilot-based OFDM systems. Simulations have confirmed the capabilities of the proposed nonlinear complex LS-SVM in the presence of Gaussian and impulse noise interfering with the pilot symbols for a high mobile speed when compared to LS standard method. The proposal takes into account the temporal-spectral relationship of the OFDM signal for highly selective channels. The Gram matrix using RBF kernel lead to a significant benefit for OFDM communications especially in those scenarios in which impulse noise and deep fading are presents.

REFERENCES

[1]   C. Lim, D. Han, "Robust LS channel estimation with phase rotation for single frequency network in OFDM," *IEEE Transactions on Consumer Electronics*, Vol. 52, pp. 1173–1178, 2006.

[2]   S. Galih, T. Adiono and A. Kumiawan, "Low Complexity MMSE Channel Estimation by Weight Matrix Elements Sampling for Downlink OFDMA Mobile WiMAX System," *International Journal of Computer Science and Network Security (IJCSNS)*, February 2010.

[3]   A. Baynast, A. Sabharwal, B. Aazhang, "Analysis of Decision-Feedback Based Broadband OFDM Systems," *Conference on Signals Systems & Computers ACSSC*, 2006.

[4]   A. Omri, R. Bouallegue, R. Hamila and M. Hasna, "Channel estimation for LTE uplink system by perceptron neural network," *International Journal of Wireless & Mobile Networks (IJWMN)*, Vol. 2, no. 3, pp. 155–165, August 2010.

[5]   S. Rahman, M. Saito, M. Okada and H. Yamamoto, "An MC-CDMA signal equalization and detection scheme based on support vector machines," *Proc. of 1st International Symposium on Wireless Communication Systems*, Vol. 1, pp. 11–15, 2004.

[6]   S. Chen, A. K. Samingan and L. Hanzo, "Support vector machine multiuser receiver for DS-CDMA signals in multipath channels," *IEEE Trans. on Neural Networks*, Vol. 12, no. 3, pp. 604–611, 2001.

[7]   M. J. Fernández-Getino García, J. M. Páez-Borrallo, and S. Zazo, "DFT-based channel estimation in 2D-pilot-symbol-aided OFDM wireless systems," *IEEE Vehicular Technology Conf.*, vol. 2, pp. 815–819, 2001.

[8]   M. Sliskovic, "Signal processing algorithm for OFDM channel with impulse noise," *IEEE conf. on Electronics, Circuits and Systems*, pp. 222–225, 2000.

[9]   J. L. Rojo-Álvarez, C. Figuera-Pozuelo, C. E. Martínez-Cruz, G. Camps-Valls, F. Alonso-Atienza, M. Martínez-Ramón, "Nonuniform Interpolation of Noisy Signals Using Support Vector Machines," *IEEE Trans. Signal process.*, vol. 55, no.48, pp. 4116–4126, 2007.

[10]   L. Nanping, Y. Yuan, X. Kewen, and Z. Zhiwei, "Study on Channel Estimation Technology in OFDM system," *IEEE Computer Society Conf.*, pp.773–776, 2009.

[11]   S. Coleri, M. Ergen and A. Puri, "Channel estimation techniques based on pilot arrangement in OFDM systems," *IEEE Trans. on broadcasting*, vol. 48, no.3, pp. 223-229, 2002.

[12]   M. Martínez Ramón, N. Xu, and C. G. Christodoulou, "Beamforming Using Support Vector Machines," *IEEE antennas and wireless propagation J.*, vol. 4, 2005.

[13]   M. J. Fernández-Getino García, J. L. Rojo-Álvarez, F. Alonso-Atienza, and M. Martínez-Ramón, "Support Vector Machines for Robust Channel Estimation in OFDM," *IEEE signal process. J.*, vol. 13, no. 7, 2006.

[14]   3rd Generation Partnership Project, "Technical Specification Group Radio Access Network: evolved Universal Terrestrial Radio Access (UTRA): Base Station (BS) radio transmission and reception," *TS 36.104*, V8.7.0, September 2009.

[15]   3rd Generation Partnership Project, "Technical Specification Group Radio Access Network: evolved Universal Terrestrial Radio Access (UTRA): Physical Channels and Modulation layer," *TS 36.211*, V8.8.0, September 2009.

[16]   3rd Generation Partnership Project, "Technical Specification Group Radio Access Network: Physical layer aspects for evolved Universal Terrestrial Radio Access (UTRA)," *TR 25.814*, V7.1.0, September 2006.

[17]   3rd Generation Partnership Project, "Technical Specification Group Radio Access Network: evolved Universal Terrestrial Radio Access (UTRA): Physical layer procedures," *TS 36.213*, V8.8.0, September 2009.

14

Improving MANET Routing Protocols through the Use of Geographical Information

Vasil Hnatyshin

Department of Computer Science, Rowan University, Glassboro, NJ, USA
hnatyshin@rowan.edu

ABSTRACT

This paper provides a summary of our research study of the location-aided routing protocols for mobile ad hoc networks (MANET). This study focuses on the issue of using geographical location information to reduce the control traffic overhead associated with the route discovery process of the ad-hoc on demand distance vector (AODV) routing protocol. AODV performs route discovery by flooding the whole network with the route request packets. This results in unnecessarily large number of control packets traveling through the network. In this paper, we introduced a new Geographical AODV (GeoAODV) protocol that relies on location information to reduce the flooding area to a portion of the network that is likely contains a path to destination. Furthermore, we also compared GeoAODV performance with that of the Location Aided Routing (LAR) protocol and examined four mechanisms for reducing the size of the flooding area: LAR zone, LAR distance, GeoAODV static, and GeoAODV rotate. We employed OPNET Modeler version 16.0 software to implement these mechanisms and to compare their performance through simulation. Collected results suggest that location-aided routing can significantly reduce the control traffic overhead during the route discovery process. The comparison study revealed that the LAR zone protocol consistently generates fewer control packets than other location-aided mechanisms. However, LAR zone relies on the assumption that location information and traveling velocities of all the nodes are readily available throughout the network, which in many network environments is unrealistic. At the same time, the GeoAODV protocols make no such assumption and dynamically distribute location information during route discovery. Furthermore, the collected results showed that the performance of the GeoAODV rotate protocol was only slightly worse than that of LAR zone. We believe that even though GeoAODV rotate does not reduce the control traffic overhead by as much as LAR zone, it can become a preferred mechanism for route discovery in MANET.

KEYWORDS

Mobile Ad-Hoc Networks; MANET Routing Protocols; Ad Hoc On Demand Distance Vector Routing; Location-Aided Routing; Geographical AODV; OPNET Modeler

1. INTRODUCTION

In this paper we summarize the results of our research endeavors in the area of location-aided routing protocols for mobile ad hoc networks (MANET). In particular, our study focuses on the issue of using geographical location information to reduce the control traffic overhead associated with the route discovery process. MANET routing protocols often rely on flooding to find a route to destination. Flooding is a simple and effective technique for route discovery where each node broadcasts route request message to all of its neighbors. The neighboring nodes repeat the process until a route to destination is found. The message broadcasts are usually "heard" by all the neighboring nodes, including those that have already "seen" (i.e., forwarded) that message. To deal with such situations, MANET routing protocols typically include a mechanism that allows the nodes to identify and discard message duplicates. However, flooding often leads to unnecessary large control message overhead since the whole network is being searched, including the parts of the network that do not contain a route to destination.

In our study we examined how the route discovery process of Ad-Hoc On-demand Distance Vector (AODV) protocol [1-3] can be enhanced through the use of geographical location information. AODV is a stateless reactive routing protocol for MANET, which employs flooding to discover a route to destination. There have been numerous studies that attempted to improve the performance of MANET route discovery through the use of geographical location information [4-14]. This study concentrates on Location-aided routing (LAR) protocol [7-8] and its improvements [5-6]. LAR is a modification of AODV protocol which relies on Global Positioning System (GPS) coordinates and traveling velocity of the nodes to limit flooding to a small area which is likely to contain a path to destination. This approach reduces the amount of control traffic traveling through the network during the route discovery process because only a portion of the network is being searched. Geographical AODV (GeoAODV) [5-6] is a variation of LAR protocol which we developed and studied over the past several years. GeoAODV also relies on the GPS coordinates to reduce the size of the search area during route discovery. However, unlike LAR, GeoAODV assumes that the nodes only know their own location information, while the GPS coordinates of all the other nodes in the network are dynamically distributed during route discovery. GeoAODV defines the search area differently from LAR and it allows expanding the area if the initial attempt to find a route to destination fails. The main contributions of this paper are (1) introduction of a new location-aided protocol called GeoAODV and (2) summary of a simulation study that compares the performance of the AODV protocol and its location-aided variations: LAR zone, LAR distance, GeoAODV static, and GeoADV rotate.

We used OPNET Modeler [15] version 16.0 to implement LAR and GeoAODV protocols and to conduct our simulation studies. OPNET Modeler is leading commercial software for simulation and modeling of computer networks. It supports a wide range of network protocols, technologies, and device models which allows the users to model almost any of today's computer networks. OPNET Modeler relies on combination of the C/C++ code, state transition diagrams, and discrete event simulation engine to model various devices, communication mediums, applications, protocols, and network technologies. While the accuracy of the results generated by OPNET products is very high [15], it comes at the expense of the underlying implementation complexity. Typically, an OPNET model of network protocols such as IP, consists of thousands and thousands lines of C/C++ code distributed through numerous external files and process model modules. Even though the implementation is mostly well documented, identifying the location of the code responsible for modeling certain aspects of the simulated system often is a very challenging task. This paper includes an overview of our endeavors implementing GeoAODV and LAR protocols using OPNET Modeler software.

The rest of the paper is organized as follows. In Section 2 we provide an overview of related work, followed by introduction of GeoAODV protocol in Section 3. We describe our implementation of LAR and GeoAODV protocols in Section 4. Description of the simulation study and analysis of collected results are presented in Sections 5 and 6, respectively. The paper concludes and presents the plans for the future work in Section 7.

2. RELATED WORK OVERVIEW

2.1. Ad-hoc On-demand Distance Vector (AODV)

As the name implies, the Ad-hoc On-demand Distance Vector (AODV) routing protocol is an on-demand protocol that begins a route discovery process only when source has data to send but does not know a route to destination. There are two main parts of the AODV protocol: route discovery and route maintenance. We are primarily interested in the route discovery phase that performs network-wide flooding to discover a path to destination. The route maintenance part of the AODV protocol deals with the removal of the outdated or broken path entries from the routing table [1-3]. However, this part of AODV has little or no influence on the route discovery process and thus is not consider in this study.

The AODV route discovery phase is conducted as follows. The sources node, also referred to as an originator, initiates the process by broadcasting the Route Request (RREQ) packet. The RREQ packet is re-broadcast further by the intermediate nodes until it reaches the destination node or a node that knows a path to destination. The request packet header carries such information as the RREQ packet ID, the originator IP address, the originator sequence number, the destination IP address, the destination sequence number, and others. As RREQ travels through the network, the intermediate nodes update/build their routing tables by recording an id of the hop from which RREQ arrived, along with the originator's IP address and sequence number. These routing table entries in the intermediate nodes form a reverse path to the originator node. The intermediate nodes also keep track of recently rebroadcast RREQ packets, by recording the originator IP address, the originator sequence number, and the RREQ ID field values. The intermediate nodes identify and discard all duplicate and outdated request packets. The duplicate packets are identified via the originator IP address and the RREQ packet ID, while outdated RREQs are identified via the originator's IP address and the sequence number [1-3].

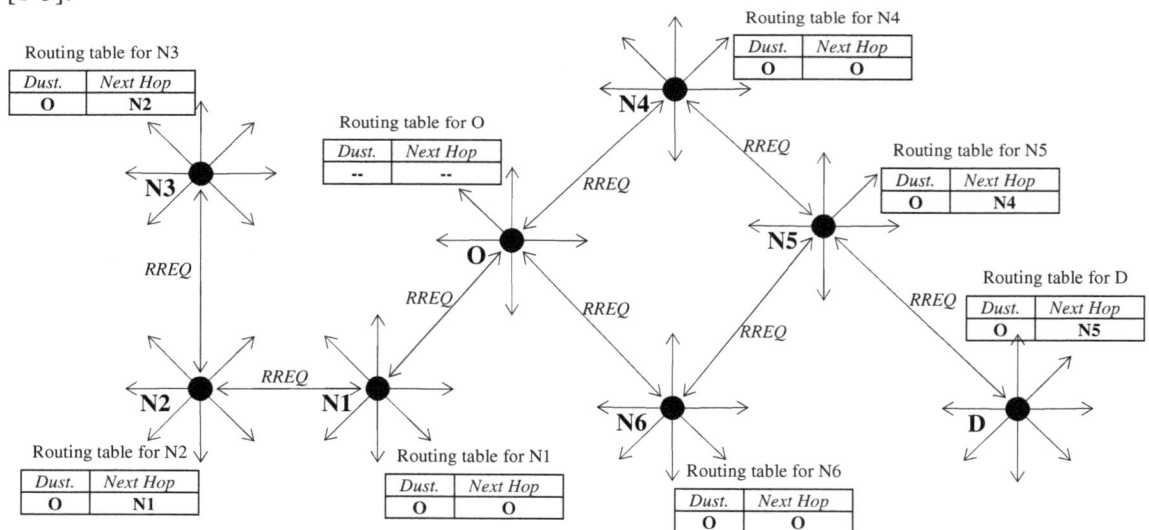

Figure 1. Establishing path to originator during the RREQ flooding

The AODV sequence numbers represent the "freshness" of information and are used to identify the outdated RREQ packets and to find "newer" routes to destination. When an intermediate node that already has a route to destination, receives the RREQ packet, it compares the destination sequence number recorded in its routing table with that carried in the RREQ packet header. If the routing table's destination sequence number is greater than that retrieved from the RREQ packet, then an intermediate node has a "fresh" route to destination, which can be advertised in the network. In this case, the intermediate node generates the Route Reply (RREP) packet and sends it to originator. Similarly, the Route Reply (RREP) packet is also sent when the RREQ packet arrives at the destination node. The RREP packets are unicast back to the originator node using the reverse route recorded during the RREQ flooding.

As RREP travels through the network, the nodes update their routing tables by recording the IP address of the node from which RREP arrived. This process creates a forward path from originator to destination. The destination sequence number carried in each RREP packet is also recorded in the routing table. The destination sequence number is used to check if another RREP that arrives at the node carries a fresher route to destination. Specifically, if the node receives the RREP packet with the value of destination sequence number higher than that recorded in its routing table then the node updates its routing table entry by replacing the destination sequence number with the value carried in the RREP packet and setting the next hop field to the id of the node that sent RREP. Route discovery completes when the originator node receives the RREP packet and starts transmitting data [1-3].

Figures 1 and 2 illustrate the AODV route discovery process in which node **O** attempts to discover a route to node **D**. As shown in Figure 1, when node **O** initiates route discovery and floods the network with the RREQ packets, all the nodes receive generated RREQ and update their routing tables. For example, when node **N2** receives the RREQ packet originated by node **O** and rebroadcasted by node **N1**, **N2** adds an entry into its routing table recording that to reach **O** the data should be forwarded to **N1**. Please note that when **N2** rebroadcasts RREQ, node **N1** will receive that RREQ but will discard it as a duplicate. When destination node **D** receives the RREQ packet it unicasts RREP back to **O**. However, since MANET is a broadcast environment, the neighbors of node **D** and of all the intermediate nodes on the path to **O** will hear the RREP packet and will update their routing tables accordingly. For example, when node **N6** (which is not part of the path between **O** and **D**) overhears the RREP message forwarded by node **N5**, it adds a routing table entry which states that to reach **D** the data has to be forwarded to **N5**.

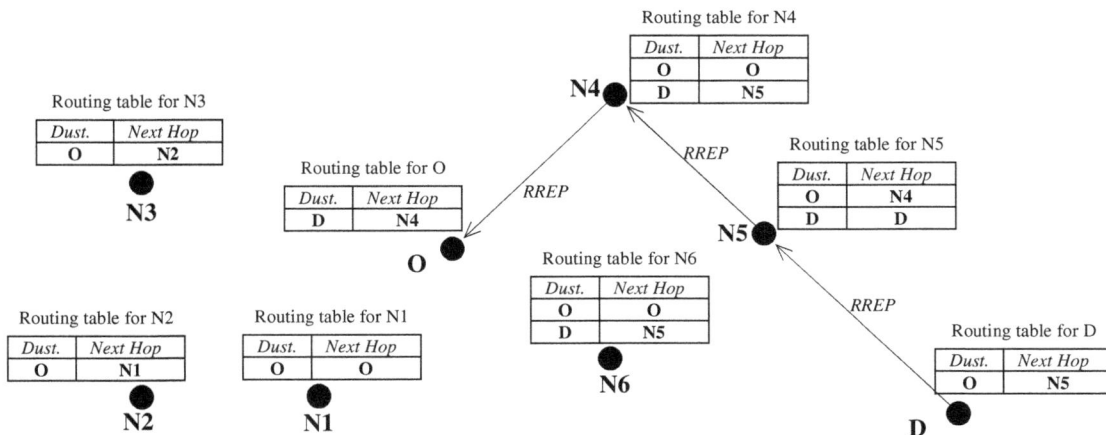

Figure 2. Establishing path to destination via RREP

To reduce the negative effects of network-wide flooding, AODV relies on the expanding ring search mechanism, where the scope of the flooding is controlled by varying the value of the Time-to-Live (TTL) field in the IP header. The originator node first attempts to find a route to destination by setting the TTL field to some initial value, typically 1. If originator does not receive a RREP message within certain amount of time then it assumes that a route to destination was not found and it repeats the process again using large TTL value. This process continues until a route to destination is found or the route discovery process with the TTL field in the RREQ packet set to a certain maximum value times-out (i.e., fails to find a route to destination) [1-3].

2.2. Location-Aided Routing (LAR)

The Location-Aided Routing (LAR) protocol [7-8] is an extension of AODV, which attempts to limit the flooding area during the route discovery process. LAR assumes that all the nodes know the Global Positioning System (GPS) locations and the travelling velocities of all the other nodes in the network. LAR relies on this information to identify a portion of the network which is likely to contain a path to destination. Only the nodes that belong to the identified part of the network participate in route discovery and rebroadcast the RREQ messages.

2.2.1 LAR Zone

There are two primary variations of the LAR protocol which we will refer to as *LAR zone* and *LAR distance*. LAR zone computes the area of the network where the destination node is likely to be located using the last know GPS coordinates and the traveling velocity of that destination node. This area is called *the expected zone* and it is defined as a circle with radius R, centered in the last-known GPS location of the destination node recorded at certain time t_0. The value of radius **R** is computed according to equation (1), where v is the last-known traveling speed of destination and t_1 is the current time:

$$R = v \times (t_1 - t_0) \tag{1}$$

Next, LAR zone computes the area, called *the request zone*, which is likely to contain a path to destination. LAR zone defines the request zone as a smallest rectangle that encompasses the expected zone so that the sides of the rectangle are parallel to the X and Y axes. During route discovery, the LAR zone protocol confines the RREQ flooding to the request zone, i.e., only the nodes within the request zone rebroadcast the RREQ messages. Figure 3 illustrates possible arrangements of the expected and request zones in the LAR zone protocol.

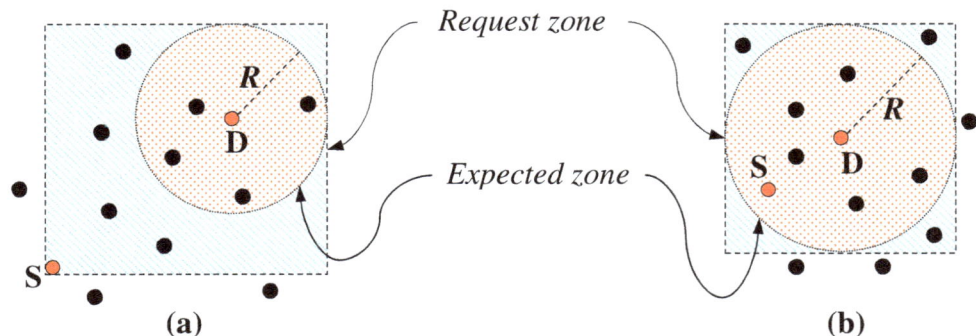

Figure 3. Arrangement of the expected and request zones of the LAR zone protocol:
(a) source is located <u>outside</u> of the destination's expected zone
(b) source is located <u>inside</u> of the destination's expected zone

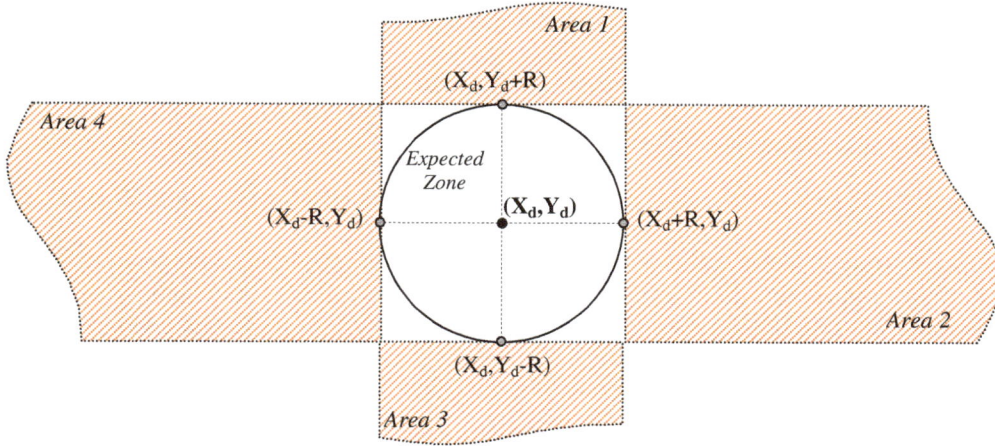

Figure 5. Special Cases for the definition of the request zone in LAR zone scheme

While studying the LAR protocol, we discovered several special cases, not cover in the literature, which pertain to the request zone definition. To simplify the explanation we use (X_s, Y_s) and (X_d, Y_d) to denote the source and destination node coordinates, respectively and R to denote the radius of the expected zone. Imagine the expected zone circle surrounded by a square with the sides of length R that parallel to X and Y axes. Now imagine that each side of that square is a part of a rectangular area that extends to infinity in the direction away from the expected zone circle. Please note that all the sides of these four rectangles are parallel to X and Y axes. Figure 5 illustrates such situation and identifies the rectangular areas as *Area 1*, *Area 2*, *Area 3*, and *Area 4*. We observed that if the source node is located in any of the rectangular areas 1 – 4, then the request zone, as defined in the LAR protocol, is be unable to cover all of the expected zone area, i.e., it is impossible to create the requested zone rectangle with the sides tangent to the expected zone circle and parallel to Y and Y axes if the source node is located in any of the rectangular areas 1 – 4. This situation occurs if the X, Y coordinates of the source node satisfy one of the following rules:

- $X_s \in (X_d - R, X_d + R)$ and $Y_s > Y_d + R$ – the source node is located in *Area 1*
- $X_s > X_d + R$ and $Y_s \in (Y_d - R, Y_d - R)$ – the source node is located in *Area 2*
- $X_s \in (X_d - R, X_d + R)$ and $Y_s < Y_d - R$ – the source node is located in *Area 3*
- $X_s < X_d - R$ and $Y_s \in (Y_d - R, Y_d - R)$ – the source node is located in *Area 4*

To resolve this issue we extended the definition of the request zone to include a portions of the expected zone that otherwise is left out. Specifically, we specified the new (X, Y) coordinates of the lower-left and upper-right corners of the redefined request zone as shown in equations (2) and (3):

$$Lower-left\ corner: (\min(X_d - R, X_s), \min(Y_d - R, Y_s)) \tag{2}$$

$$Upper-right\ corner: (\max(X_d + R, X_s), \max(Y_d + R, Y_s)) \tag{3}$$

2.2.2 LAR Distance

The LAR distance variation of the LAR protocol does not require the knowledge of the node velocities as it does not compute the request zone area. Instead, LAR distance only assumes that the GPS location of the destination node is known network-wide and then relies on this information to limit the RREQ rebroadcast to only those nodes that are closer to destination than the node which forwarded the RREQ packet [7-8]. This rule is generalized through equation 4, where α and β are configuration parameters, $\mathbf{D}$ is the destination node, $\mathbf{N_0}$ is the node that broadcasts RREQ, $\mathbf{N_1}$ is the node that receives RREQ from $\mathbf{N_0}$, and $|\mathbf{A\ B}|$ denotes the distance between nodes $\mathbf{A}$ and $\mathbf{B}$.

$$\alpha \times |N_0 D| + \beta \geq |N_1 D| \tag{4}$$

Figure 6 illustrates an example of the LAR distance protocol operation. Suppose the source node $\mathbf{S}$ initiates the route discovery process by broadcasting the RREQ packet. At some point, node $\mathbf{N_0}$ receives this RREQ and rebroadcasts it farther. When node $\mathbf{N_1}$ receives RREQ from node $\mathbf{N_0}$ it participates in route discovery and rebroadcasts the packet again since $\mathbf{N_1}$ is located closer to destination than node $\mathbf{N_0}$, i.e., $|\mathbf{N_1\ D}| \leq |\mathbf{N_0\ D}|$. However, nodes $\mathbf{N_2}$ and $\mathbf{N_3}$ discard the RREQ packet received from $\mathbf{N_0}$ because both of these nodes are located father away from destination $\mathbf{D}$ than node $\mathbf{N_0}$, i.e., $|\mathbf{N_2\ D}| > |\mathbf{N_0\ D}|$ and $|\mathbf{N_3\ D}| > |\mathbf{N_0\ D}|$.

Even though the LAR protocols achieve the goal of reducing the control traffic overhead during the route discovery process, they both suffer from two deficiencies. First, the LAR protocols assume a network-wide availability of GPS coordinates, which often is not the case. Second, both protocols do not account for a possibility that a route to destination cannot be found even though it does exist, i.e., in the case of LAR zone the route could be located outside the request zone and in the case of LAR distance a portion of the route could require traveling away from destination. Our approach, called Geographical AODV, attempts to address these issues by only assuming that the nodes know their own GPS coordinates (the coordinates of other nodes in the network are dynamically distributed during the route discovery process) and by expanding the request zone area and re-starting the route discovery process again if a path to destination was not found using a smaller request zone area.

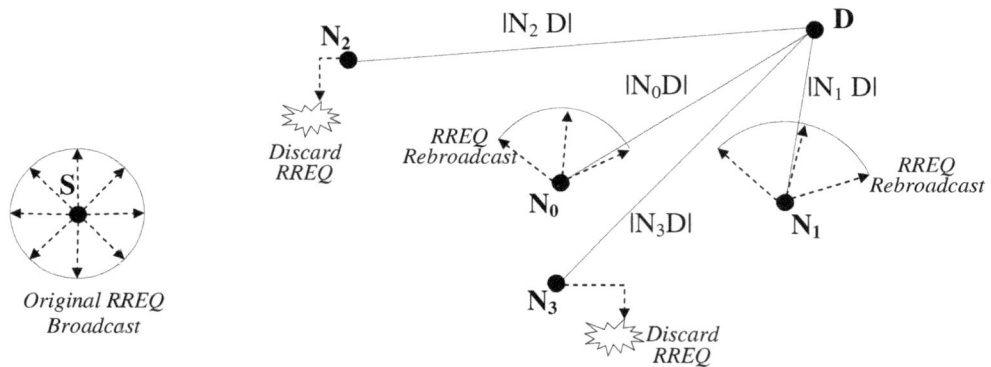

Figure 6. Example of the LAR distance protocol operation

3. GEOGRAPHICAL AODV (GEOAODV)

Geographical AODV (GeoAODV) [4-6] is our proposed improvement of LAR zone protocol: it defines the request zone differently than LAR zone, it allows the request zone area to expand after route discovery failures, and it dynamically distributes the node coordinates during route discovery. In this section we introduce the GeoAODV protocol and describe its operation in detail.

3.1 GeoAODV Protocol

GeoAODV defines the request zone in a shape of an isosceles triangle, where the source node is the vertex of the triangle, located in the top corner opposite to the base (i.e., the source node is the origin point of the equal sides of the triangle). The destination node is located on the line that originates at the source node and is perpendicular to the base of the triangle. The width of the GeoAODV request zone (i.e., the isosceles triangle) is controlled via the angle between the equal sides. We refer to this protocol configuration parameter as *the flooding angle* and denote it as α. Only the nodes located within the confines of the GeoAODV request zone participate in route discovery. All the other nodes discard arriving RREQs. Intermediate node **N** computes angle θ formed between the source node, itself, and destination (as shown in Figure 7) to determine if it belongs to the request zone. Since the source-destination vector always divides flooding angle evenly, node **N** belongs to the request zone if angle θ is not larger than one half of flooding angle α.

$$\theta \le \frac{1}{2} \times \alpha \tag{5}$$

Thus, if inequality (5) holds then node **N** is located within the request zone and will rebroadcast the RREQ packet, otherwise **N** is outside of the request zone and RREQ is discarded. Generally we compute the value of angle θ according to equation (6), where we use $\overrightarrow{SD}$ to denote a vector between source node **S** and destination node **D**, $\overrightarrow{SN}$ to denote vector between source node **S** and node **N**, while $|SD|$ and $|SN|$ are the absolute values of vectors $\overrightarrow{SD}$ and $\overrightarrow{SN}$, respectively.

$$\theta = \cos^{-1}\left(\frac{\overrightarrow{SD} \cdot \overrightarrow{SN}}{|SD| \times |SN|}\right) \tag{6}$$

At the start of the route discovery process GeoAODV sets flooding angle α to some initial value. This initial value could be determined by the "freshness" of the destination's GPS coordinates, i.e., the value of α increases proportionally to t_d, the time passed since the last update of the destination's location information. Once t_d crosses certain threshold, α is set to 360 degrees and GeoAODV performs the same way as regular AODV. Alternatively, the initial value of the flooding angle could be a function of the expected zone radius defined in the equation (1).

If an attempt to find a route to destination using certain value of the flooding angle fails (i.e., a time out occurs) then GeoAODV repeats route discovery again using a larger value of the flooding angle, which effectively expands the size of the request zone. GeoAODV continues this process until a path to destination is found or until GeoAODV searched the whole network (i.e., route discovery with flooding angle value of 360 degrees failed to find a route to destination). Please note that when the flooding angle value reaches 360 degrees, GeoAODV operates exactly the same way like the AODV protocol. Since GeoAODV may eventually search the whole network, it guarantees that a route to destination will be found if one exists [5-6].

Figure 7 illustrates an example of the GeoAODV protocol operation where source node **S** initiates the route discovery process in an attempt to find a path to destination node **D**. Initially, **S** uses the flooding angle with the value α_1. The request zone defined by α_1 is shown in Figure 7 as an isosceles triangle of a lighter grey color. During this round of route discovery only intermediate node N_1 rebroadcasts the RREQ packets, while the remaining nodes are outside the request zone defined by α_1 and are excluded from participating in route discovery. These nodes (i.e., N_2, N_3, and N_4) discard all arriving RREQs during this initial round of route discovery. If the first round of route discovery fails, then source increases the flooding angle to some new value α_2 and repeats the process again. During the second round of route discovery, the request zone is extended (i.e., shown in Figure 7 as a darker color isosceles triangle) and intermediate nodes N_1, N_2, and N_3 rebroadcast RREQs, while intermediate node N_4 discards all arriving RREQ packets since it is located outside the request zone defined by flooding angle α_2.

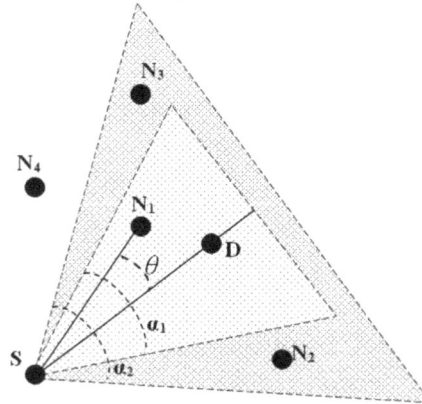

Figure 7. Example of GeoAODV operation

3.2 GeoAODV Rotate

In our study we considered two variations of the GeoAODV protocol: *GeoAODV static* and *GeoAODV rotate*. GeoAODV static operates as described above: the request zone remains unchanged during each round of route discovery, i.e., the source node is always a vertex opposite to the base of the isosceles triangle. GeoAODV rotate operates a bit differently. It re-orients the request zone towards destination at each intermediate node by making the previous node a new vertex of the triangle, i.e., each intermediate node re-computes the request zone based on the location of the previous hop, instead of the source node. Figure 8 illustrates the idea of GeoAODV rotate: node N_1 belongs to the request zone computed based on location of node S while node N_2 belongs to the new, re-oriented request zone computed based on the location of node N_1. Both nodes N_1 and N_2 participate in route discovery even though they belong to different request zones. On the other hand, node N_3, which receives RREQ from N_1, will not participate in route discovery because node N_3 is located outside the new request zone computed based on the location of N_1, its previous hop. On the other hand, when GeoAODV static is used, all the node in the Figure 8 will participate in route discovery because they all belong to the request zone computed based on location of source node S.

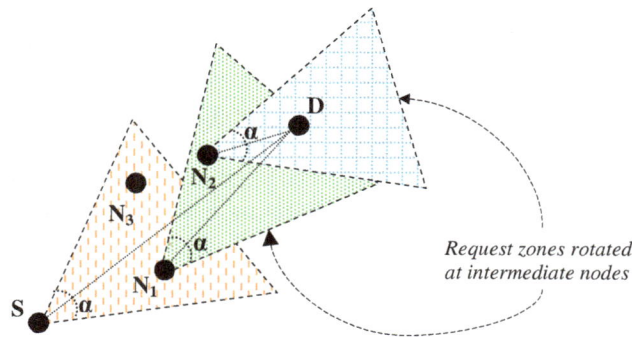

Figure 8. Example of GeoAODV rotate operation

3.3 Distributing Location Information in GeoAODV

GeoAODV assumes that the nodes use GPS to find their own coordinates, while the locations of all the other nodes is dynamically distributed in the network during route discovery. Each GeoAODV node stores the node location data in a structure we call *the geo-table*. An entry in the geo-table consists of the location information (e.g., GPS coordinates), the "freshness" timer, the AODV sequence number, and the identity of the node (e.g., IP address). The "freshness" timer is keeps track of when the node coordinates were updated the last time, while the AODV sequence number allows an intermediate node to identify if the arriving control packet (e.g., RREQ or RREP) carries new location information. This process is similar to the way the AODV protocol differentiates between new and old control packets. However, the geo-table entries remain valid for longer periods of time than the entries in the AODV routing table because the location information can help determine the general direction where the destination node may be located even if a route to that destination has changed [5-6].

We modified format of the RREQ and RREP packets to carry additional information such as locations of the source and destination nodes, and the flooding angle. This information is used to populate the geo-table and to determine if an intermediate node should participate in route discovery. At the start of the route discovery process the source node consults it geo-table and generates a RREQ packet that will carry node's own location information, the initial value of the flooding angle, and the last known location of the destination node. If the source node does not contain a geo-table entry for the destination node then the flooding angle is set to 360 degrees and GeoAODV operates the same way as regular AODV.

Upon the RREQ message arrival, the nodes (even those that will discard the RREQ packet) update their geo-tables with the source node's location information. An intermediate node updates its geo-table with the destination's location information only if the destination sequence number carried in RREQ is larger than that stored in its geo-table. Otherwise, an intermediate node may update the destination coordinates carried in the RREQ packet. Similar processing occurs when RREP is sent back: each intermediate node updates its geo-table with the source and destination location information carried in the packet. GeoAODV also utilizes periodic AODV hello messages (which have the same header format as the RREQ packets) to distribute location information among the neighboring nodes [5-6].

4. GEOAODV IMPLEMENTATION IN OPNET MODELER

We used OPNET Modeler [15] version 16.0 to implement the LAR and GeoAODV protocols. OPNET Modeler is the leading industrial software for modeling computer networks. It employs layered architecture which consists of the process, node, network, and simulation domains. The process domain allows the developer to model behavior of various processes such as applications, network protocols, physical environments, etc. This domain includes such items as external C/C++ code files and various process models. A process model consists of the finite state machine that describes the modeled process, implementation of various actions using C or C++ programming language, and various configuration parameters. At the node domain layer individual process models are combined together to create the node models of various networking devices such as servers, switches, routers, WLAN devices, etc. Figure 9 shows OPNET node model of MANET station [16].

The network domain combines individual node and link models to create and configure representation of simulated network. At the network domain level the developer can specify various characteristics of the network such as the types, configuration, and location of individual nodes and links, the traffic generation sources, the traveling speed of individual nodes, parameter settings for various network protocols, etc. The simulation domain is responsible for configuring simulation model as a whole and for specifying the values of various parameters such as the duration of the simulation, statistics to be collected during the simulation, the seed value for the random number generator, etc.

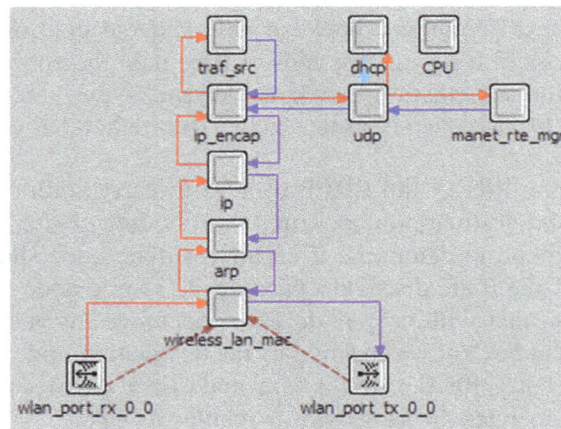

Figure 9. OPNET node model of MANET station

To implement the LAR and GeoAODV protocols we modified OPNET *aodv_rte* process model and several external C files including *manet_support*, *aodv_suport*, *aodv_pkt_support*, *aodv_request_table*, *aodv_route_table*, and *aodv_packet_queue* all of which are responsible for modeling the AODV protocol. Figure 10 illustrates *aodv_rte* process model. The finite state machine representation of the AODV protocol consists of two states: *init* which initializes state variables, configuration parameters, supporting data structures, etc., and *wait* state which models operation of the AODV protocol. Wait state is responsible for processing various arriving AODV control packets, generating new AODV control packets and preparing them departure, forwarding and queuing (if necessary) the data packets, and generating periodic AODV hello messages. Upon the packet arrival the process model identifies the type of an incoming packet (e.g., data, RREQ, RREP, etc) and then calls the corresponding function that handles processing of the packets of identified type.

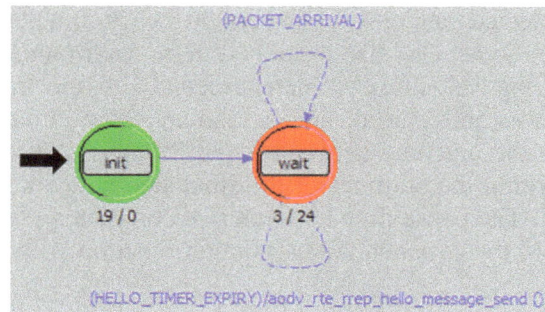

Figure 10. OPNET Modeler's *aodv_rte* process model

When modifying OPNET's implementation of the AODV protocol, we first added several parameters to simplify configuration of the LAR and GeoAODV protocols. In OPNET Modeler, configuration parameters for MANET routing protocols are defined via *manet_mgr* process model. However, the parameter values are parsed in the corresponding routing protocol process models, such as *aodv_rte*. Next, we modified the structure of the RREQ and RREP headers, defined in *aodv_pkt_support.ex.h* file, to carry additional information. After that we created several new data structures to store location information. We created a simulation-wide table, which we refer to as *global_location_table*, which helps us to model the network-wide availability of geographical coordinates and velocities of the nodes in the LAR protocol. This table is periodically updated by individual nodes that store their precise location information and traveling speed. The frequency of these periodic updates is controlled via configuration parameter which by default is set to 1 second. During the LAR route discovery process individual nodes consult this table to retrieve location information about other nodes in the network. We also implemented the geo-table which stores such information as the IP address, the insertion time, location coordinates, and the sequence number needed for the GeoAODV protocol. The information stored in a node's geo-table is only accessible by that node itself and is not available network-wide.

Finally we modified *aodv_rte* process model and several external files to implement the LAR and GeoAODV route discovery processes. Specifically, we added the code responsible for
- processing arriving control packets,
- generating the RREQ, RREP, and HELLO messages using new GeoAODV format,
- determining if a node should retransmit an RREQ packet based on LAR zone, LAR distance, GeoOADV static, or GeoADOV rotate approach, and
- updating location information in *global_location_table* and *geo-table* data structures.

We verified the accuracy of our implementation by conducting a line-by-line code trace under various configuration settings.

5. SIMULATION SET-UP

We compared the performance of the AODV, LAR, and GeoAODV protocols in the network environment that consisted of 50 MANET node randomly placed within a 1500 meters x 1500 meters network domain area. In this study we varied the following two parameters: the number of nodes that generate traffic and the node traveling velocities. Specifically, we conducted simulation studies with 2, 5, 10, 20, and 30 communicating nodes, where all the source-destination pairs were selected randomly. Furthermore, each simulation set with a different number of communication nodes was also examined in the environment where the nodes were traveling with the following velocities: 0 meters/second (all the nodes were stationary), 5 meters/second, 10 meters/seconds speed, and a random value computed using the uniform distribution function with the outcome between 0 and 20 meters/second. All the nodes in the network were traveling within the confines of the simulated network domain according to the Random Waypoint model. The pause time between two consecutive legs of the node's journey was computed according to the exponential distribution function with the mean outcome of 10 seconds.

TABLE 1. Summary of Node Configuration

Configuration Parameter	Value
Channel Data Rate	11 Mbps
Transmit Power	0.0005 Watts
Packet Reception Power Threshold	-95 dBm
Start of data transmission	normal (100, 5) seconds
End of data transmission	End of simulation
Duration of Simulation	300 seconds
Packet inter-arrival time	exponential (1) second
Packet size	exponential (1024) bytes
Mobility model	Random Waypoint
Pause Time	exponential(10)
Destination	Random

We configured the LAR protocol to have the nodes publish their location and traveling velocities in *global_location_table* once every second. We set the values of configuration parameters α and β for the LAR distance protocol to 1 and 0, respectively. Furthermore, we modified the LAR protocols to operate like regular AODV if they fail to find a path to destination after the initial round of route discovery, i.e., if the LAR protocol does not find a route to destination then it tries again using the AODV protocol. We configured the GeoAODV protocol to have the initial value of the flooding angle set to 90 degrees if the destination coordinates are known and to 360 otherwise. The value of the flooding angle was incremented by 90 degrees after each failed attempt to find a route to destination. The wireless LAN configuration parameters of each node were set to their default values as defined in OPNET Modeler. The summary of simulation set-up is provided in Table I.

We executed a total of 100 different simulation scenarios (i.e., 5 different values for the number of communicating nodes and 4 different traveling velocity values). Each simulation scenario was executed for 300 seconds with the communicating nodes starting data transmission at simulation time of 100 seconds. Each scenario was executed six times using different seed value for the random number generator. We executed a total of 600 simulation runs, which took over 72 hours to complete. Collected results from this study were exported into a comma separated text file and then processed using a Python programming language script.

6. RESULTS

It should be noted that in this simulation study we made few simplifying assumptions. Specifically, we did not account for the delay associated for the retrieval of GPS coordinates; in our evaluation of the GeoAODV protocols we disregarded the overhead introduces by addition of new fields in the RREQ and RREP headers; and in the LAR protocols we made an assumption that location information and traveling velocities are available everywhere in the network at no additional cost. Our study primarily focused on the total amount of control traffic generated by each of the examined protocols. The results of this study suggest that all location-aided routing protocols outperform AODV by generating significantly fewer control packets during route discovery. The summary of collected results is in presented in Figures 11 – 15 which illustrate the total number of control packets (i.e., the number of RREP + RREP) generated by each protocol versus the node traveling speed.

The results show that the LAR zone protocol consistently generates the smallest number of control packets while GeoAODV rotate is a close second. This can be attributed to the fact that the simulation does not account for the cost associated with retrieval of node coordinates and traveling speeds in LAR zone (i.e., these values are assumed to be available as needed). GeoAODV on the other hand makes no such assumption, and distributes location information during route discovery. Additionally, before location-aided improvements of GeoAODV kick-in, the nodes require some time to gather location information about other nodes in the network. As a result, initially, GeoAODV operates the same way as regular AODV, and only after location information has been distributed in the network it can take advantage of the limited flooding. Furthermore, GeoAODV may have to go through up-to 3 rounds of route discovery using different values of the flooding angle before conducting route discovery using the AODV protocol. The LAR protocols, on the other hand, revert to AODV after a single failure.

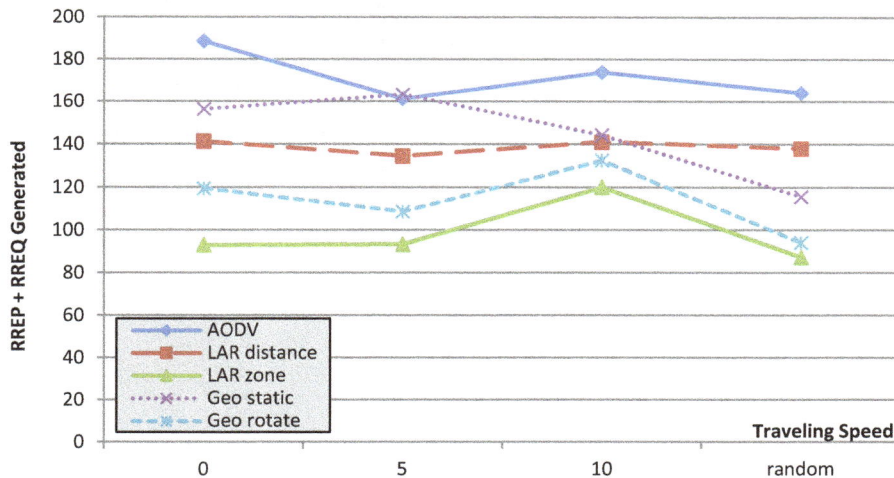

Figure 11. The control traffic overhead in scenarios with 2 communicating nodes

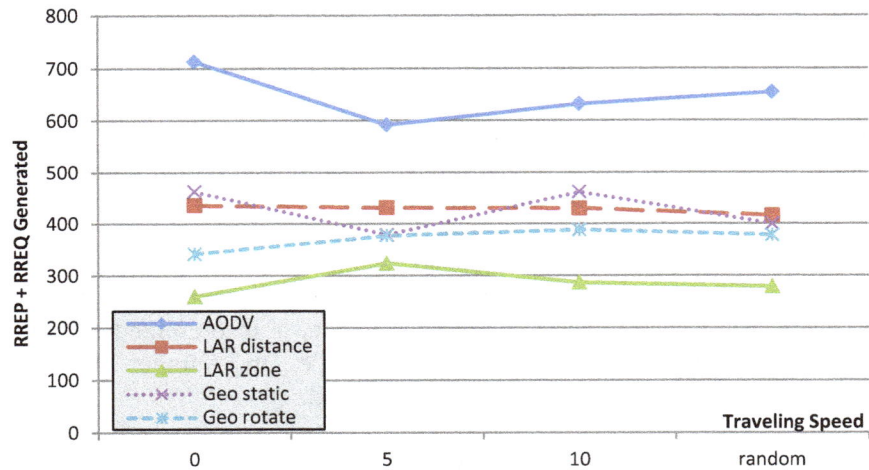

Figure 12. The control traffic overhead in scenarios with 5 communicating nodes

The difference in performance between two LAR protocols can be attributed to how often the protocols do not find a route and are forced to conduct route discovery as AODV. The search area of the LAR distance protocol is very limited: the current node rebroadcasts the RREQ message only if it is closer to destination than the previous node. Thus, it is not surprising that LAR distance fails to find a route to destination more frequently than LAR zone. As result, LAR distance often behaves like AODV and generates a large number of control packets. LAR zone, on the other hand, conducts route discovery over a wide area and thus is less likely to revert to AODV. This results in LAR zone consistently outperforming the LAR distance protocol.

When comparing the GeoAODV protocols we observed that GeoAODV rotate consistently outperforms GeoAODV static. While both variations of GeoAODV fail to find a route to destination roughly the same number of times, GeoAODV rotate dynamically reorients the direction of the request zone and thus excludes the nodes that likely are not a part of a route to destination. This results in GeoAODV rotate forwarding fewer RREQ packets through the network and thus introducing lower control traffic overhead than the GeoAODV static protocol.

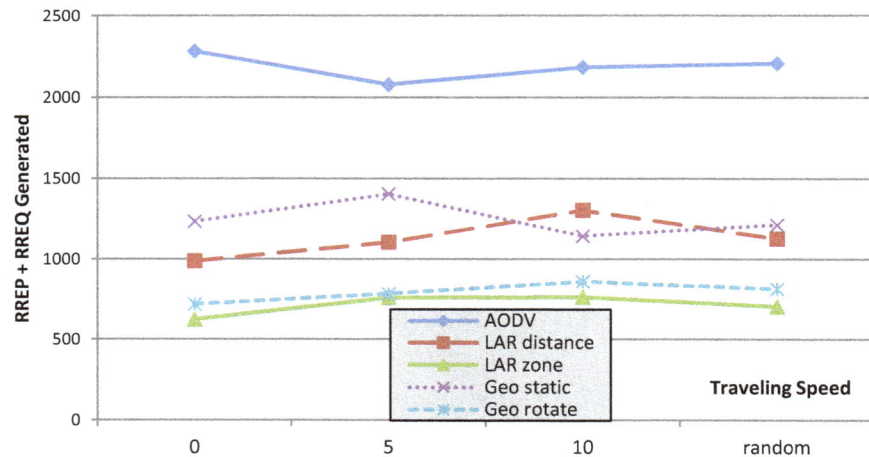

Figure 13. The The control traffic overhead in scenarios with 10 communicating nodes

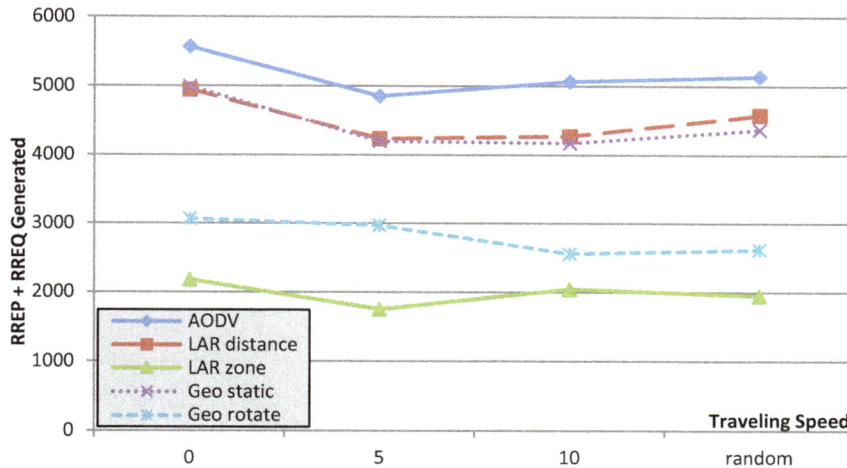

Figure 14. The control traffic overhead in scenarios with 20 communicating nodes

As expected, collected results report that the control traffic overhead increases with the increase in the number of communicating nodes. What was surprising is how all the protocols, except for LAR zone, performed in the simulation scenario with 30 communication nodes. As Figure 15 shows, LAR distance, GeoAODV static, and GeoAODV rotate generated almost the same amount of control traffic as AODV. Such behavior could be attributed to the fact what when there are many communicating nodes, the chance of failing to find a route using limited broadcast increase causing these protocols to revert to regular AODV more frequently. As a result, all of the advantage gained by successfully employing limited flooding is lost when the protocols fail to find a route and have to conduct a network-wide flooding. LAR zone appears to be less susceptible to this problem and performs the best as shown in Figure 15. Nevertheless, the GeoAODV rotate protocol consistently remains a second-best option, outperforming all the protocols except for LAR zone in all evaluated scenarios.

7. CONCLUSIONS AND FUTURE WORK

This paper introduced a new routing protocol called GeoAODV and presented comprehensive comparison study of the AODV-based location-aided routing protocols. Simulation results suggest that even though GeoAODV rotate does not always reduce the control traffic overhead by as much as LAR zone, it can become a preferred mechanism for route discovery in MANET. The main advantages of GeoAODV are its build-in mechanism for distributing location information and its simplicity of deployment. LAR zone requires the location coordinates and the traveling speed of individual nodes to be readily available in the network, which may not be always possible. However, the LAR zone protocol has an advantage in the environments where location information and node traveling speed is distributed with the help of some additional facilities such as described in [17-21].

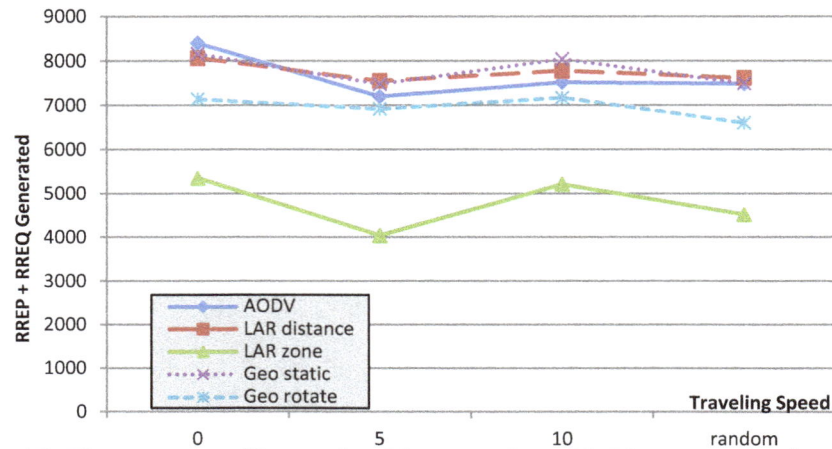

Figure 15. The control traffic overhead in scenarios with 30 communicating nodes

Currently we continue our study focusing on other aspects of protocol performance such as the number of route discovery failures and the time to find a route to destination. We are also studying the accuracy of location information available in the network during GeoAODV route discovery and how that accuracy influences the protocol performance. We are investigating other possible improvements of the GeoAODV protocol including selection of initial value for the flooding angle, dynamically adjusting the flooding angle at intermediate nodes (i.e., increasing the flooding angle value when an intermediate node knows that there are no neighboring nodes within the request zone defined by the flooding angle carried in arriving RREQ), and few others. We are also examining improvements to the LAR protocols which will allow them to increase the search area after the route discovery failure, instead of immediately reverting to AODV. For example, LAR distance can adjust the values of configuration parameters α and β used in (4), while LAR zone can increase the request zone by "advertising" the source node coordinates as if source is located farther away on the line between source and destination than it actually is. Finally, we plan to expand our study by comparing the performance of LAR and GeoAODV with other routing protocols, such as the Greedy Perimeter Stateless Routing (GPSR) protocol [22] and the Geographical Routing Protocol (GRP) [23].

ACKNOWLEDGEMENTS

I would like to thank Hristo Asenov, Remo Cocco, Malik Ahmed, Dan Urbano, Rob Hussey, Earl Huff, and Zabih Shinwari, the students from Rowan University, Computer Science Department, who have been tirelessly working with me on this project over the past several years. Thank you all for your dedication, diligence, and hard work!

REFERENCES

[1]   C. Perkins, E. Belding-Royer, S. Das. (July 2003). *Ad hoc On-Demand Distance Vector (AODV) Routing*. IETF RFC 3561. Last accessed on 2013-01-26

[2]   E. M. Royer and C. E. Perkins. *An Implementation Study of the AODV Routing Protocol*, Proc. of the IEEE Wireless Communications and Networking Conference, Chicago, IL, September 2000

[3] C. E. Perkins and E. M. Royer. *Ad hoc On-Demand Distance Vector Routing*, Proc. of the 2nd IEEE Workshop on Mobile Computing Systems and Applications, New Orleans, LA, Feb. 1999, pp. 90-100

[4] V. Hnatyshin, M. Ahmed, R. Cocco, and D. Urbano, *A Comparative Study of Location Aided Routing Protocols for MANET*, Proc. of 4th IEEE IFIP Wireless Days 2011 conference, Niagara Falls, Canada

[5] V. Hnatyshin, and H. Asenov, *Design and Implementation of an OPNET model for simulating GeoAODV MANET routing protocol*, Proc. of the OPNETWORK 2010 International Conference, Session: Wireless Ad Hoc and Wireless Personal Area Networks, Washington DC, August 2010

[6] H. Asenov, and V. Hnatyshin, *GPS-Enhanced AODV routing*, Proc. of the 2009 International Conference on Wireless Networks (ICWN'09), Las Vegas, Nevada, USA (July 13-16, 2009)

[7] Y. Ko and N. H. Vaidya, *Location-aided routing (LAR) in mobile ad hoc networks*, Wireless Networks, 6(4), July 2000, pp. 307-321

[8] Y. Ko and N. H. Vaidya, *Flooding-based geocasting protocols for mobile ad hoc networks*, Mobile Networks and Applications, 7(6), Dec. 2002, pp. 471-480

[9] A. Husain, B. Kumar, A. Doegar, *A Study of Location-Aided Routing (LAR) Protocol for Vehicular Ad Hoc Networks in Highway Scenario*, International Journal of Engineering and Information Technology, 2(2), 2010, pp. 118-124

[10] K.M.E. Defrawy and G. Tsudik, *ALARM: Anonymous Location-Aided Routing in Suspicious MANETs*, Proc. of the IEEE International Conference on Network Protocols, 2007, pp. 304-313

[11] D. Deb, S. B. Roy, N. Chaki, *LACBER: A new Location-Aided routing protocol for GPS scarce MANET*, International Journal of Wireless & Mobile Networks (IJWMN), 1(1), August 2009

[12] F. De Rango, A. Iera, A. Molinaro, S. Marano, *A modified location-aided routing protocol for the reduction of control overhead in ad-hoc wireless networks*, Proc. of the 10th International Conference on Telecommunications, 2003, pp. 1033 – 1037

[13] Y. Xue, B. Li, *A Location-aided Power-aware Routing Protocol in Mobile Ad Hoc Networks*, Proc. of the IEEE Global Telecommunications Conference, San Antonio, TX, November 2001, pp. 2837 – 2841

[14] Y. Wang, L. Dong, T. Liang, X. Yang, X., D. Zhang, *Cluster based location-aided routing protocol for large scale mobile ad hoc networks*, IEICE Transactions, 2009, E92-D(5), pp. 1103–1124

[15] OPNET Modeler ver. 16.0. OPNET Technologies, Inc®, www.opnet.com last visited 2/1/13

[16] Adarshpal S. Sethi and Vasil Y. Hnatyshin, *The Practical OPNET User Guide for Computer Network Simulation*, 527 pages, Chapman and Hall/CRC (August 24, 2012), ISBN-10: 1439812055, ISBN-13: 978-1439812051

[17] H. Cheng, J, Cao, H. Chen and H. Zhang, *GrLS: Group-Based Location Service in Mobile Ad Hoc Networks*, IEEE Transactions on Vehicular Technology, Vol. 57, No. 6, November 2008

[18] Gajurel, S., Heiferling, M., *A Distributed Location Service for MANET Using Swarm Intelligence*, Mobile WiMAX Symposium, 2009. MWS '09. IEEE, pp. 220 – 225

[19] Yongming Xie, Guojun Wang, Jie Wu, *Fuzzy Location Service for Mobile Ad Hoc Networks*, Computer and Information Technology (CIT), 2010 IEEE 10th International Conference on, pp. 289 – 296

[20] S. M. Das , H. Pucha and Y. C. Hu *Performance comparison of scalable location services for geographic ad hoc routing*, Proc. INFOCOM 2005, vol. 2, pp.1228 -1239

[21]   K. Wolfgang , F. Holger and W. Jorg *Hierarchical location service for mobile ad hoc networks*, ACM SIGMOBILE Mobile Comput. Commun. Rev. 2004, vol. 8, no. 4, pp. 47 – 58

[22]   B. Karp and H. T. Kung, "GPSR: greedy perimeter stateless routing for wireless networks," in MobiCom '00: Proceedings of the 6th annual international conference on Mobile computing and networking. New York, NY, USA: ACM Press, August 2000, pp. 243–254.

[23]   OPNET Modeler 16.1 Documentation, OPNET Technologies, Inc., 2012.

15

Pedestrain Monitoring System using Wi-Fi Technology And RSSI Based Localization

[1]Zhuliang Xu, [2]K. Sandrasegaran,[3]Xiaoying Kong, [4]Xinning Zhu ,[5]Jingbin Zhao, [6]Bin Hu and [7]Cheng-Chung Lin

Faculty of Engineering and Information Technology
University of Technology Sydney
Sydney, Australia

[1]Zhuliang.Xu@student.uts.edu.au
[[2]Kumbesan.Sandrasegaran,[3]Xiaoying.Kong, [4]Xinning.Zhu]@uts.edu.au
[[5]Jingbin.Zhao, [6]Bin.Hu-1]@student.uts.edu.au
[7]Cheng-Chung.Lin@eng.uts.edu.au

ABSTRACT

This paper presentsa new simple mobile tracking system based on IEEE802.11 wireless signal detection, which can be used for analyzingthe movement of pedestrian traffic. Wi-Fi packets emitted by Wi-Fi enabled smartphones are received at a monitoring station and these packets contain date, time, MAC address, and other information. The packets are received at a number of stations, distributed throughout the monitoring zone, which can measure the received signal strength. Based on the location of stations and data collected at the stations, the movement of pedestrian traffic can be analyzed. This information can be used to improve the services, such as better bus schedule time and better pavement design. In addition, this paper presents a signal strength based localization method.

KEYWORDS

Mobile Tracking, Localization, Wi-Fi; Captured Packets, Channel Hopping, RSSI, EMD, EEMD

1. INTRODUCTION

Wi-Fi networks have been widely deployed in homes, enterprises and organizations. Wi-Fi technology is defined in various IEEE 802.11 standards (including 802.11a, 802.11b, 802.11g, and 802.11n). It is a popular method to provide Internet access for wireless users. Nowadays, smartphone has become a common device and important part of everyone's daily life. Most people carry smartphones during working, shopping and leisure time.Asmartphone can be identified by using its unique IDs like International Mobile Equipment Identity (IMEI) number or Media Access Control (MAC) address of the handset. The IMEI is sent once when a mobile registers with a network, whereas the MAC address is on every data packet sent by Wi-Fi enabled mobile handset. Each MAC frame includes destination and source MAC addresses.

Wi-Fi MAC address can be used to identify a mobile device and it can be used to determine the location of a mobile device when it is combined with received signal strength at multiple locations. A good applicationis to monitor patients in a hospital [1] or in location sensing [2]. However, there are some problems when Wi-Fi positioning is applied in outdoor conditions. A few localization methods can be applied for outdoor. The use of wireless positioning technologies have been discussed by several researchers in the past few years [3] and the most common localization approach is using received signal strength (RSS). Specifically, one method uses a propagation model to covert RSS to the distance and applied triangulation method to determine the location of the transmitter.

Another method is an empirical model which uses fingerprinting method. This method uses some RSS measured at a number of points within an area as reference point (RP). Thereafter, ituses RP to compare with the measured RSS of wireless device to estimate the location. There are a number of problems with the finger printing method (a) a general wireless device such as Wi-Fi adapter provides receiver signal strength indicator (RSSI) not RSS,(b)finger printing method requires a static outdoor environment ,(c) It has been proven that the RSSI cannot be reliably used for localization [4], due to inconsistent behaviour and the error in measured RSSI value increases as distance increases.

The idea of using Wi-Fi enabled smartphone to monitor pedestrian traffic or individual movement has been discussed in the literature in recent years. Inside a building, the pedestrian monitoring can be easily achieved by using the Wi-Fi enabled smartphone and access point (AP) [4, 5]. All the smartphones have to communicate with an AP to obtain a Wi-Fi connection. During the communication process, the AP extracts the information which is needed for smartphone tracking, for instance, the MAC address and RSSI. Therefore, the pedestrian traffic in an area of interest can be investigated by using the information. However, the smartphone based pedestrian monitoring has a lot of challenges in an outdoor environment. Most of the traffic monitoring systems deployed today uses some special purpose sensors such as magnetic loop [6], cameras [7] and RFID tag-reader. These methods are applicable with vehicles instead of pedestrians and very costly.

This paper introduces a new solution for the problems mentioned above by proposing a mobile tracking system. This system captures MAC layer information of a smartphone by using wireless sniffing and uses a method of RSSI based localization to implement positioning. The purposes of this system are pedestrian traffic monitoring and people density monitoring based on smartphone tracking in a street. This information can be used to improve the service provided to people such as better bus schedule time and better pavement design.

In the rest of this paper, Section II describes the system structure and the challenges. Section III discusses the solution for the challenges. Section IV introduces a RSSI based localization method and Section V presents and analyzes the test results. The conclusion summarizes the contributions of this paper in Section VI.

2. Methodology

2.1 System Structure

This tracking system consists ofa sniffing block and an administration block.The system structure is depicted in Figure 1. The sniffing station is used for capturing and processing packets from Wi-Fi channels. Database and tracking server are two components in the administration block which stores the processed packets from sniffing stations..

Sniffing station: It contains Wi-Fi antenna, adapter, processor, 3G module and local database. The Wi-Fi adapter uses the Wi-Fi antenna to monitor channels and capture packets sent by mobile devices. These captured packets will be sent to the processor, and then the processor reads the packets, filters out useful information such as MAC address. The data is stored in local database as a backup and sent to next block via 3G module simultaneously.

Tracking server: It contains a server and a database. This block receives the data sent from sniffing block and stores them in a database. The tracking server also allows users access to the database for acquiring information.

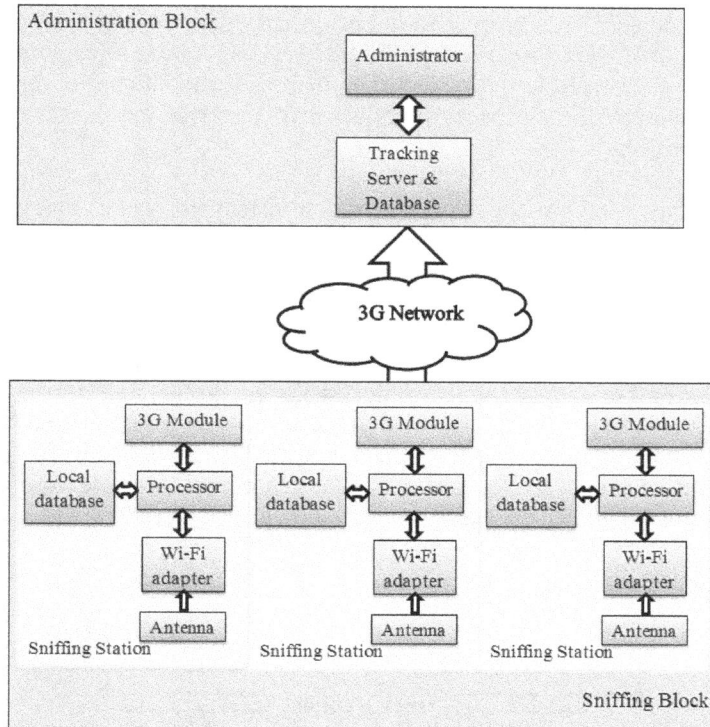

Figure 1: System structure

3G network is used as an interface between sniffing station and tracking server in this system. Using 3G networkdeliver the collected data to the tracking server can improve the coverage and availability of this system. As 3G network is widely deployed in cities, it can provide a reliable wireless communication in any places, for instant, in a park.

In addition, to configure or modify the sniffing station on site would be inconvenient in most of the situations. This system can conduct remote control and monitor the sniffing stations easily via 3G connection.

In order to reduce the amount of data transmitted through 3G interface, the data need to be filtered to onlyinclude:

- Date: date and time when the packet is captured.
- Station: station number where packet is detected.
- Device type: Wi-Fi or Bluetooth (optional)
- MAC address: The MAC address of the mobile device from which packets originate.
- Signal strength (dBm): Received signal strength from the mobile device.

2.2 Packets Sniffing

A Wi-Fi enabled smartphone in a street sends packets to discover available Wi-Fi network intermittently. The typical packet involved in this discovery process is a probe request which contains the MAC address. Therefore, system performance can be defined as device detection (amount of captured MAC addresses) and packets detection (amount of captured probe request). Forasmuch, the object allocated to the system is to maximize the number of device detected and number of packets detected. This paper describes two methods for packets sniffing: passive sniffing and active sniffing.

Using passive sniffing method, the sniffing station listens to the channel only instead of establishing communication with smartphone by sending packets. The sniffing station extracts the MAC address from each captured packet and measures the received signal strength.

The active sniffing method is achieved by using probe response injection. According to IEEE802.11 standard, Wi-Fi station (Wi-Fi enabled smartphone) communicates with the access point which is shown inFigure 2. After the station sends the probe request, it waits for the probe response in a certain time. Once the station received the probe response, it sends an acknowledgement back to the access point. The acknowledgement contains the sending station's MAC address which is necessary for tracking. The probe response injection can increase the packet sent by smartphone which is benefit to the mobile tracking system.

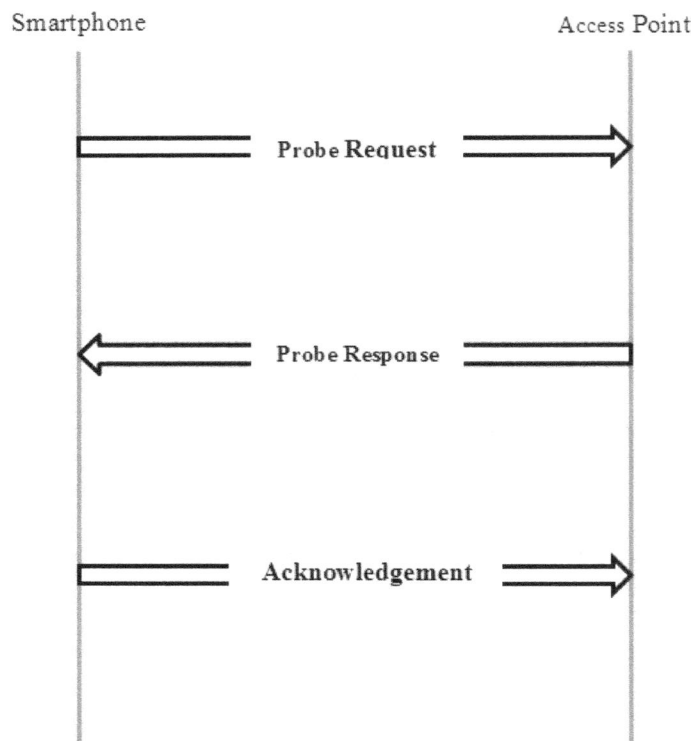

Figure 2: The probe request processing

2.3 System Challenge

This mobile tracking system facesa number of challenges. By our system design, the sniffing stationis mountedin the street to detect smartphones. However, the real world conditionsare complex. There are vehicleson the road and pedestrians with different walking speeds as they enter a building or board on a bus. Wi-Fi enabled handsets send packet slackly and randomly in the street. Therefore, this system requires a mechanism to increase the packets captured efficiently and measured RSSI accurately.

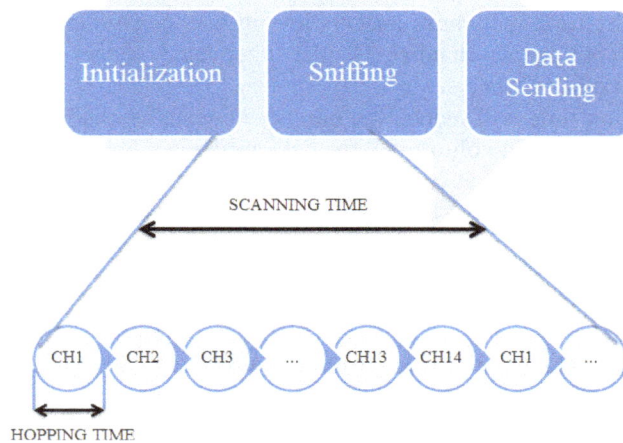

Figure 3: Hopping Time vs. Scanning Time

The radio spectrum used for 802.11 is divided into several channels, such as 802.11b/g separating the 2.4GHz spectrum into 14 channels spaced 5MHz each and Wi-Fi adapter can only operate on one frequency channel at any time. Therefore, channel hopping or frequency hopping will be used in the adapter. In order to monitor all the channel traffic without losing any packet in sniffing area, 14 measurement devices are required. Obviously it increases sniffing difficulty and cost. Therefore, channel hopping is used in this paper as a solution. This system chooses a simple packet loss avoiding method, which is achieved by using scanning time Ts and hopping time Th. As shown in Figure 3, one detecting circle is divided to sniffing phase and data sending phase. In the sniffing phase, the adapters hop on all selected channels and in this scanning period all the received signal strength values for one specific MAC address will be calculated to an average result. Scanning time and hopping time in Figure 3are the duration of one sniffing phase and the duration of adapter collecting data from one channel respectively.

Specifically, hopping time will define the hopping frequency which can tell how many channels can be scanned in one sniffing phase (i.e. scanning time). Scanning time will define result output frequency.
The relation between scanning time and hopping time is shown as the equations below

$$\frac{T_S}{T_H} = N_C (1)$$

$$N_p = \frac{f_i}{N_C} \times T \times N_{MAC} \quad (2)$$

$$N_P = \frac{f_i \times T_H}{T_S} \times T \times N_{MAC} \quad (3)$$

where N_c is the number of hopped channels in one scanning time period, N_p is the total number of captured packets, f_i is the packet sending frequency for one unique MAC address, (this frequency value is dynamically based on the network status and data demand), T is the total sniffing time and N_{MAC} is the total number of MAC addresses which stay in the sniffing area during a sniffing period.

There is a trade-off between hopping time and scanning time:

- Shorter hopping time will cause packets loss in one specific channel. Because it spends less time to monitor one channel. On the other hand, longer hopping time will cause packets loss in other channels.
- Shorter scanning time will reduce the number of scanned channels. But longer scanning time will reduce the data volume and the data accuracy which shows the possible location in a movement trace, because pedestrian is moving and the system calculates all the captured packets into one average result only within a scanning time.

2.4 Test Bed

The sniffing stations were mounted on George St. which is one of the main streets in Sydney CBD asshown inTable 1.The table also describes the Wi-Fi device which is installed in sniffing station. According to the collected data, thesmartphone involved in the tests include all majorbrands in market. (Apple, HTC, Samsung, Sony, Sharp, Blackberry, Nokia and etc.).

Table 1: The deployment of sniffing stations (Google Map)

Antenna Gain	Sensitivity
2dBi	-93dBm
Deployment	

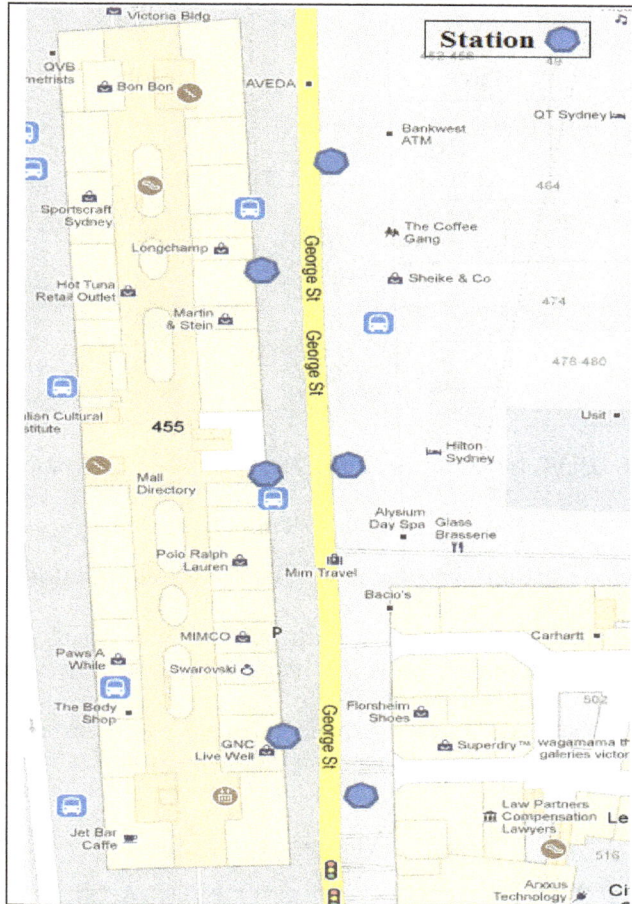

3. TEST RESULT AND ANALYSIS
3.1 The Probe Response Injection

Figure 4 demonstrated the comparison between passive and active sniffing. Thedata are collectedfor one day in the street. From the figure, we found that the active sniffing increases the number of captured packets for around 10% of observed. In this result, the contribution of active sniffing   is   fewer   than   what   it   has   been   expected.

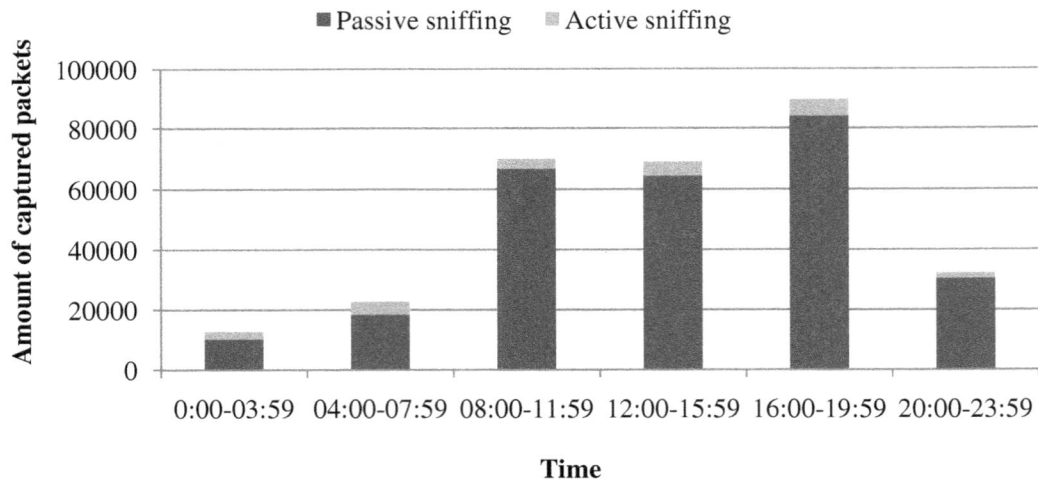

Figure 4: The comparison between passive and active sniffing

Two reasons can explain the few number packets. First, the smart phone does not reply acknowledgment for power saving which is configured by operation system. A further test was carried for different smart phone operation system. The result is presented inTable 2. Based on the result, the Apple phones and Blackberry phone never reply with acknowledgment.   Second, the smartphone and sniffing station is not tuned to the same channel, so that the probe response cannot be received by smartphone.

Table 2: The smartphone responses

Brands	Operation System	Sniffing	Packet Injection
HTC Corporation	Android	YES	YES
Apple Inc.	IOS	YES	NO
Sony Mobile	Android	YES	YES
Samsung Corporation	Android	YES	YES
Blackberry	Blackberry OS	YES	NO

3.2 Hopping Time and Scanning Time Selection

During these tests, scanning time was set to 1s, 3s and 5s. Under each scanning time the hopping time was set to 10ms, 50ms, 100ms, 250ms and 500ms and all the 14 channels were selected to scan.

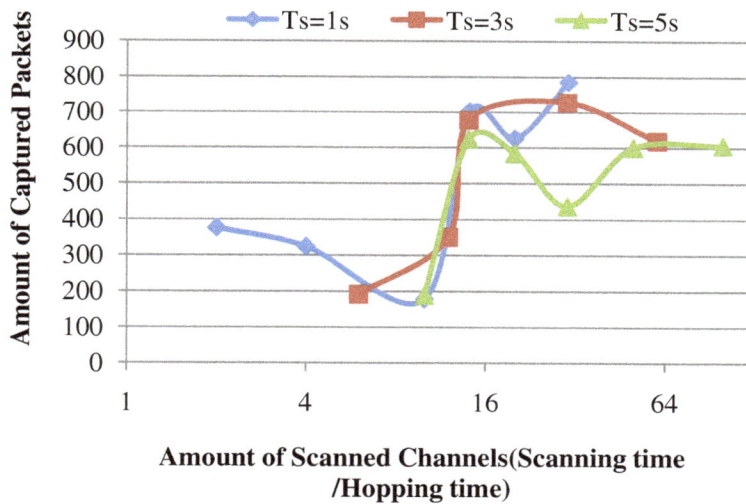

Figure 5: Scanning Time VS. Hopping Time

The result of the tests shows when the ratio between scanning time and hopping time (Eq. 3) is around multiplenumbers of selected channels, the system can capture the highest number of packets.The channel selection for packets transmitting is highly depends on the environment and the connection state. Therefore, it is hard to predict packet distribution on the channels. In this case, the time is evenly allocated to each channel for packets sniffing. This is a balance for the trade-off which is mentioned above.

Table 3: Test parameter selection

Scanning Time (s)	Hopping Time (ms)	Ratio	Hopping Time (ms)	Ratio
1	33	30.3	71	14.08
3	100	30	214	14.01
5	166	30.1	357	14.01

The graphs in Figure 5 illustrate test results in Table 3. It shows that when the ratio is around 14, the system captured largest number of packets.

Table4: Parameter selection for fixed test duration

Scanning Time (s)	Hopping Time (ms)	Ratio(Ts/Th)	Test duration (min)
1	71	14.08	2
2	142	14.08	4
3	214	14.01	6
4	285	14.03	8
5	357	14.01	10
6	428	14.01	12

Then, follow the parameters in Table 4 to fix the scanning time and hopping time ratio to 14. The corresponding test duration is used to guarantee the number of result output times is same for different setting. The result is shown in Figure 6.

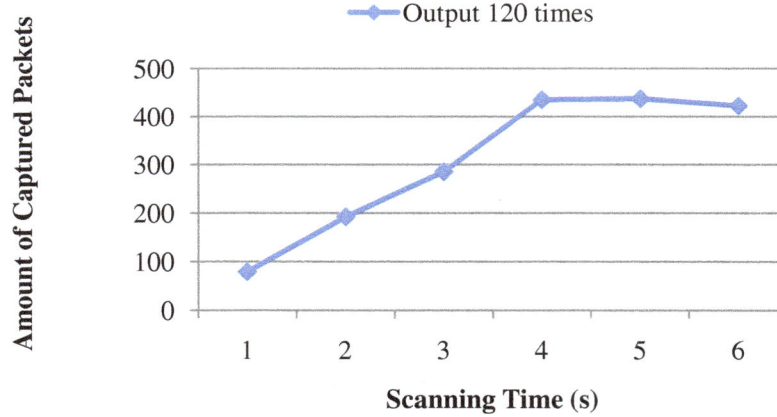

Figure 6: Fixed output amount testing result

In these tests, the scanning time does not affect the results, because the number of output is fixed. In other words, the amount of captured packets isonly affected by hopping time. Figure 6 shows that there is a limitation when hopping time is between 285ms and 357ms. A longer hopping helps sniffing station to capture more packets from one channel but lose packets from other channels. Therefore, there is a balance point between packet capturing and loss. Based on the test result, T_s=4s and T_h=285ms is the best collocation for 14 channels case.

4. RSSI Based Localization Method

Received Signal Strength Indicator (RSSI) is a parameter representing the power of received radio signal. It is measured by wireless end equipment's antenna. . However, there is not a clear relationship between RSSI and received signal power level or received signal strength (RSS).
The wireless end equipment measures the signal power level RSS and then converts this analog result to digital number which is RSSI. During the converting processing, the Analog to Digital Convertor (ADC) decides to choose the reference voltage and converting algorithm. Therefore, the same RSSI value present different RSS in different equipment.

Some related works [8, 9] present that the RSSI approach to the ten times the logarithm of the ratio between the received power (P_r) at the receiving point and the reference power (P_{ref}). Based on propagation mode, the received power is inversely proportional to the square of distance (d). The equations below summarize this relationship:

$$RSSI \propto -10\log \left(\frac{P_r}{P_{ref}}\right) \tag{4}$$

$$RSSI \propto -10\log \left(\frac{1}{d^2 \times P_{ref}}\right) \tag{5}$$

Then, using k and α to represent all uncertain factors in Eq.5, the equation can be simplified to:

$$RSSI = -klog(d) - \alpha \tag{6}$$

Figure 7 demonstrates an ideal relationship between RSSI and distance. A linear curve makes the RSSI-based positioning possible if the factors k and α wasdetermined.

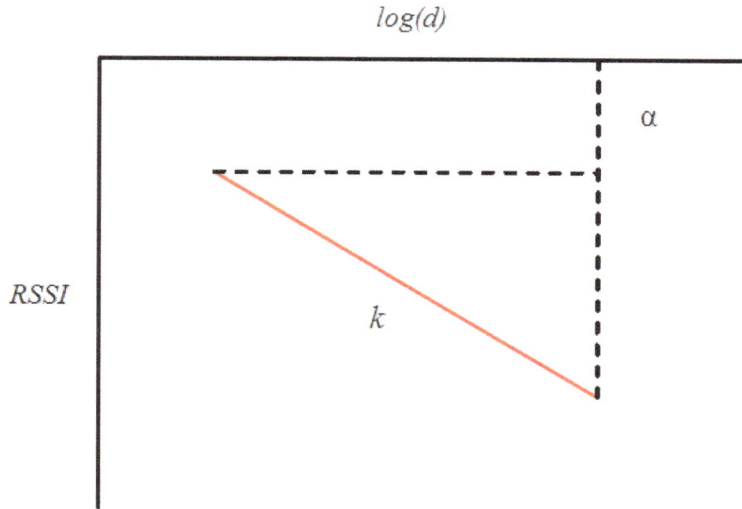

Figure 7: Ideal relationship between RSSI and distance

However, outdoor environment is dynamic and the radio signal is affected by many factors in this kind of environment such as slow fading and fast fading. Therefore, it seems to be hard to find a realistic curve.

Empirical Mode Decomposition (EMD) is an adaptive signal processing method which is highly efficient to analyze complex and non-linear signal. This method decomposes a signal (S_n) into a series of Intrinsic Mode Functions (*IMFs*) and one residue (r_n), which can be represented as the following equation [10]:

$$S_n = \sum_{k=1}^{N} imf_{k,n} + r_n \tag{7}$$

As EMD decomposes signal based on signal's own characteristics, each IMF contains different oscillation feature which is displayed as frequency from original signal. In other words, each IMF has different frequency from each other. Therefore, the noise can be filtered by selecting related IMF to reconstruct signal.

An improved EMD method is called Ensemble Empirical Mode Decomposition (EEMD) which adds white noise into original signal before using EMD to process. Some research works show that when the added white noise distributed evenly it can offset the noise in the signal [11]. The process of EEMD is:

- Add a white noise series to the signal;
- Decompose the data with added white noise into IMFs;
- Repeat step 1 and step 2, but with different white noise series
- Obtain the means of corresponding IMFs of the decompositions as the final result.

The signal strength in wireless system is affected by many factors such as: path loss, shadowing and fading. The shadowing and fading is the natural behaviour of amplitude changes in high frequency and low frequency respectively. In this case, the EMD/EEMD can be applied to remove those signal changes from the received signal and retrieve the distance related data. The process demonstrated in Figure 8.

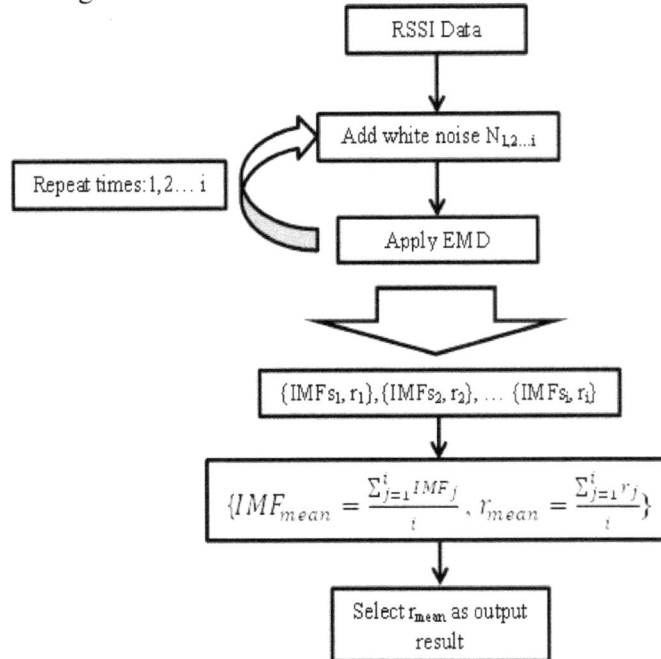

Figure 8 : The data processing using EMD method

5. LOCALIZATION

Two experiments were set up in different outdoor conditions to determine the factors k and α in Eq.6. The first Experiment A was carried out in a street at midnight which is quiet enough to avoid pedestrians' and vehicles' effects with one sniffing stations. During this experiment, 27 points in a straight line with 1m gap were measured and at each distance. 50 signal strengths were collected, and then the EMD/EEMD with i equals to 500 is used to process each point. The result is shown in Figure 9 below.

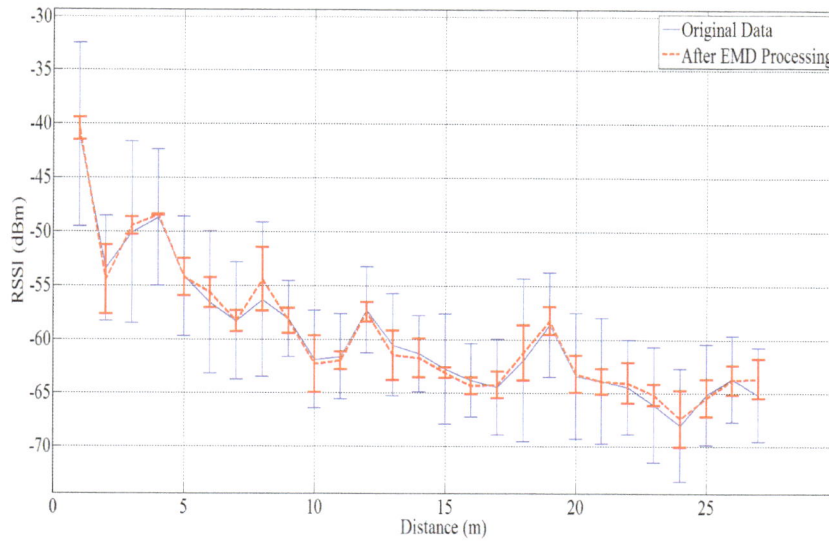

Figure9: Experiment A result

In **Figure9 Figure**, the solid line connects the mean values of each point's data and the dash line connects the mean value of processed data at each point. It is obviously that, the blue line (data) has a larger error or standard deviation and after using EMD method, the mean value does not change too much but the error is reduced. Besides, the data energy (received signal strength) does not change during adding white noise and keeping the residue only as result. Therefore, the EMD/EEMD method does not distort signal which shows this method not only reduce the oscillation but also increase the accuracy.

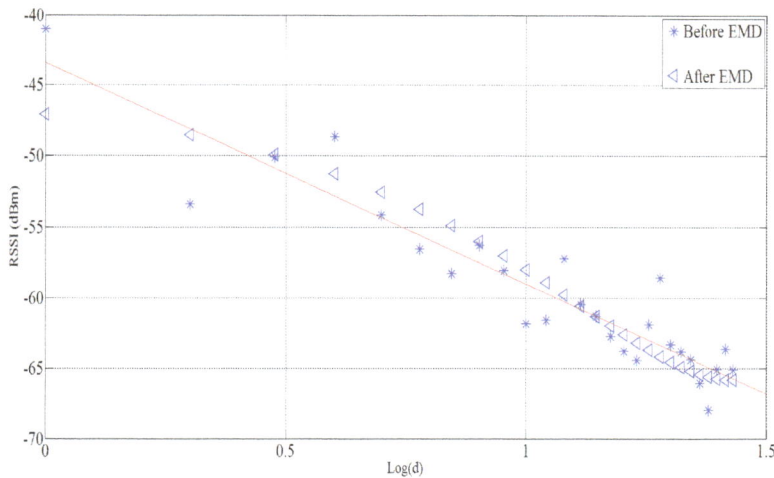

Figure10: RSSI vsLog(d) in Experiment A

Figure 10 demonstrates the relationship between RSSI and logarithm of distance. The star marks present the mean value of original data, the triangle marks are the results of applying EMD/EEMD to the mean value of original data and the line is the optimal fitting line for these data. From the figure, it can be seen that the triangle marks in Figure 10 are closer to the fitting line, which means these data will provide smaller error when calculating the distance. The fitting line shows an equation which is:

$$RSSI = -15.31 \log(d) - 41.05 \quad (8)$$

where 15.31 and 41.05 arethe value of factors k and α.

The Experiment B was carried out on the street with high traffic load and high signal interference. And repeat the steps in Experiment A. The result is shown in **Figure 11**.

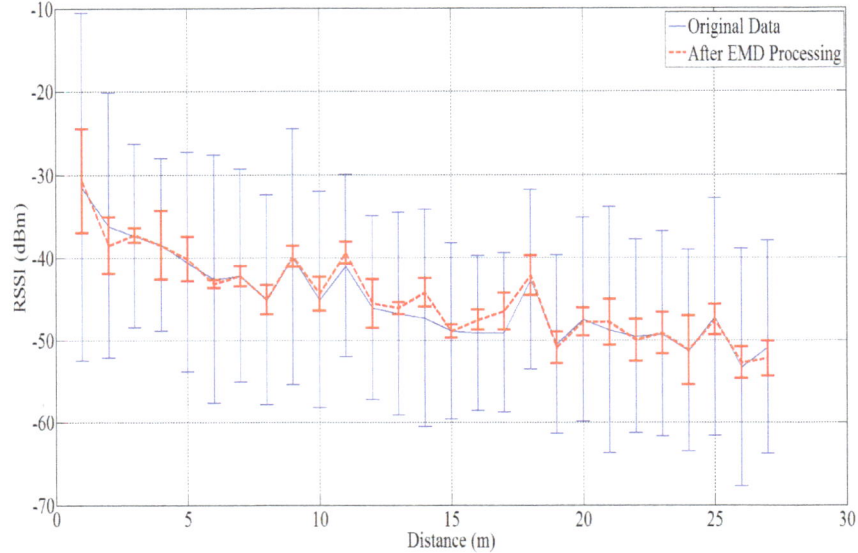

Figure 11: Experiment B result

Due to the dynamic environment condition, a larger standard deviation than Experiment A is shown in **Figure 11**. However, the standard deviation is reduced significantly after applying EMD/EEMD method.

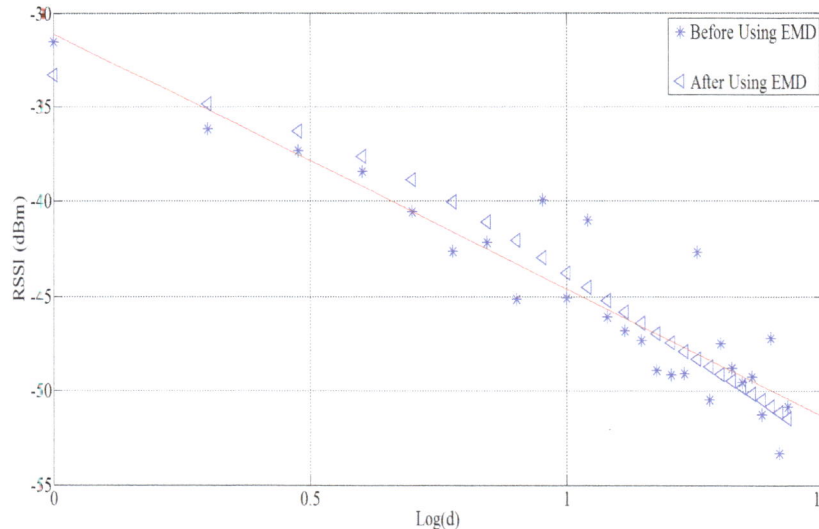

Figure 12: RSSI vsLog(d) in Experiment B

Similar to Experiment A, the line in **Figure 12** is the fitting line for these data. It follows the equation below:

$$RSSI = -14.85log(d) - 35.45 (9)$$

where 14.85 and 35.45 arethe value of factors k and α.

Experiment A and B arethe two extreme situations for an urban condition. Hence, the results from these two experiments can be regarded as upper and lower boundary. To obtain the average value of two results and set the factor k to 15.08 and the factor α to 38.45. Therefore, the relationship between RSSI and distance is:

$$RSSI = -15.08 \log(d) - 38.45 \ (10)$$

In order to examine Eq.10, Experiment C was carried out with the setting which is shown in Figure 13. The line is the movement path of mobile device and the points are the sniffing stations' locations.

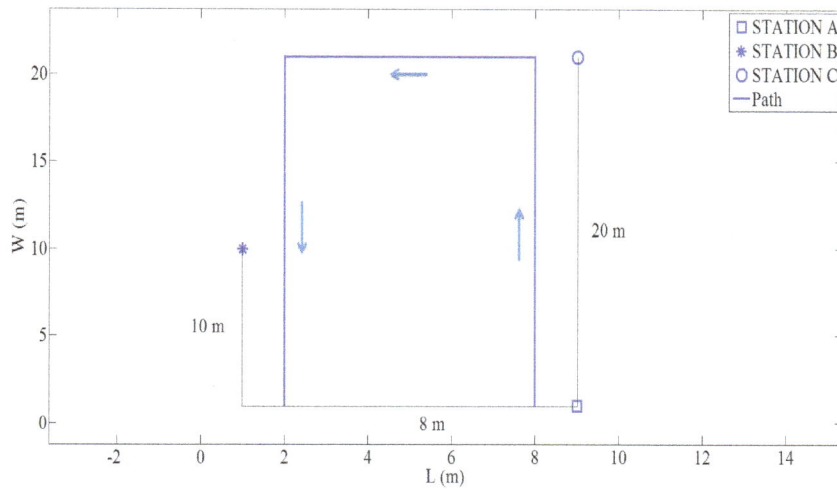

Figure 13: Experiment C set up

Figure 14 presents the collected data in test run C. The points present the original data collected by each station and the lines present the processed data by using EMD/EEMD. The results from station A and station C match with the movement path.

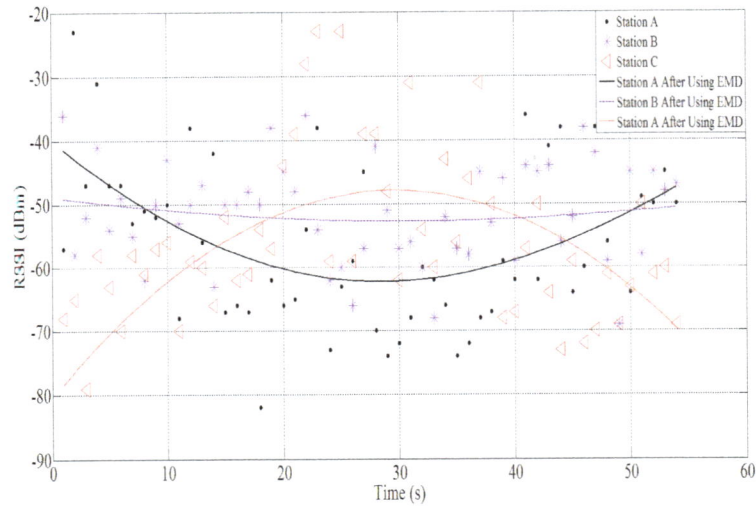

Figure 14: Experiment C result

Then, we use triangulation method to do localization based on the data collected by these three stations.

1. Use the station position as center, calculated distance as the radius to draw a circle.
2. The intersection point of three circles is the location.
3. If there are more than one or no intersection point, the center of the smallest triangle can be constructed as the location.

The result is demonstrated in **Figure 15**. The dash line presents the reverted path by using Eq.10 and processed data. The solid line is the real path. The movement can be observed by comparing the solid line and dash line.

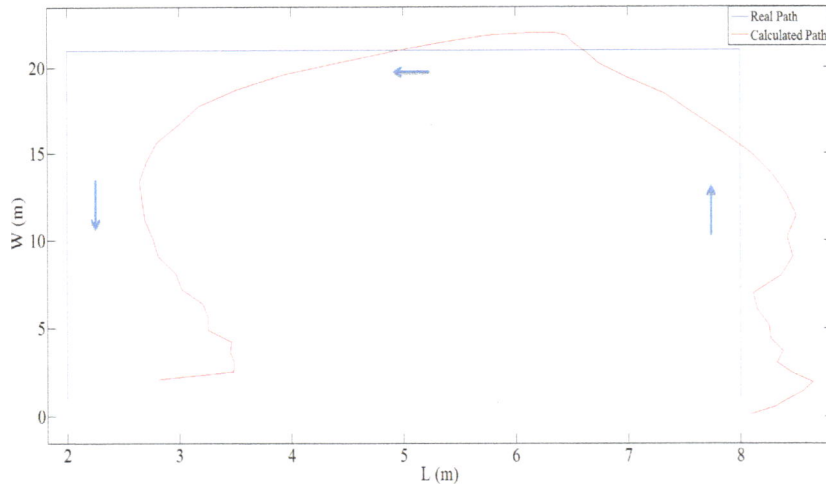

Figure 15: Normalized localization result

The position calculation error is shown in Figure 16 with maximum error of 3.13m and minimum error of 0.08m and average error of 0.998m. This error is caused by the selection of factors k and α, and the environment interference. The selected factor in this paper is an average value for different environment quality in the investigation area. It presents a general relationship between

RSSI and distance. In other hand, some distortion still remains in the RSSI signal after processing.

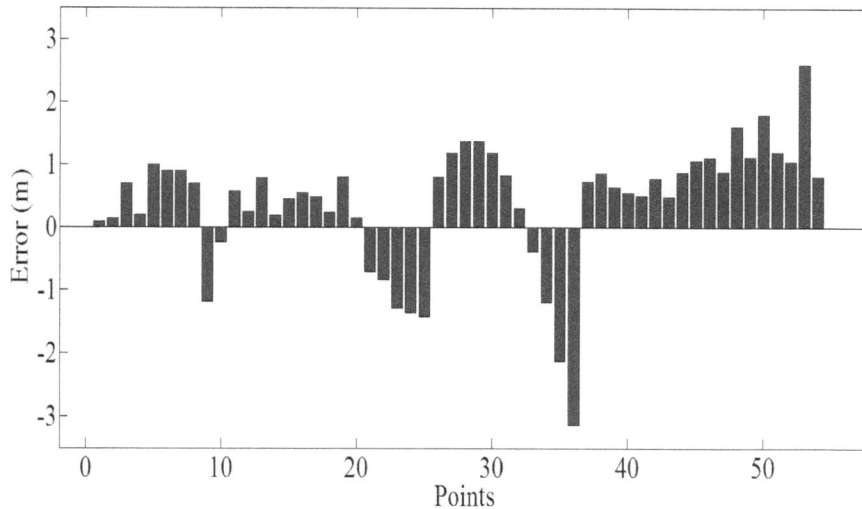

Figure 16: Localization error

6. CONCLUSION

In this paper, an outdoor large-scale mobile tracking system and a data processing method arepresented. A pair of scanning and hopping time is discovered to avoid packet loss and increase measurement accuracy.

After a series of experiments, the EMD/EEMD method has been proved that it is suitable for RSSI based localization, as it can remove the noise effect from original signal. Based on the positioning result, the mobile tracking system can be used to detect the wireless user movement with medium-precision localization with 0.998m average error. In future research thismobile tracking system will be deployed in streets using a mesh network to achieve sniffing station data exchange and centralized data collection.

REFERENCES

[1]   L. YanXia, C. Liting, and S. Yuqiu, "Application of WiFi Communication in Mobile Monitor," in *Information Engineering and Electronic Commerce, 2009. IEEC '09. International Symposium on*, 2009, pp. 313-317.

[2]   U. Bandara, M. Hasegawa, M. Inoue, H. Morikawa, and T. Aoyama, "Design and implementation of a Bluetooth signal strength based location sensing system," in *Radio and Wireless Conference, 2004 IEEE*, 2004, pp. 319-322.

[3]   A.Bensky. Wireless positioning technologies and applications. Artech House, Inc., 2007.

[4]   Z. Vasileios, B. Honary, and M. Darnell. "Indoor 802.1 x based location determination and real-time tracking." Wireless, Mobile and Multimedia Networks, 2006 IET International Conference on. IET, 2006.

[5]   S. Ronni, and T. Berglund. "Location tracking on smartphone using IEEE802. 11b/g based WLAN infrastructure at ITU of Copenhagen."

[6]   Ndoye, Mandoye. "Vehicle detector signature processing and vehicle reidentification for travel time estimation." Transportation Research Board 87th Annual Meeting. No. 08-0497. 2008.

[7]   Anagnostopoulos, C-NE. "License plate recognition from still images and video sequences: A survey." Intelligent Transportation Systems, IEEE Transactions on 9.3 (2008): 377-391.

[8]   P. AmbiliThottam, M. Iftekhar Husain, and S.Upadhyaya. "Is RSSI a reliable parameter in sensor localization algorithms-an experimental study." Proceedings of the field failure data analysis workshop, Niagara Falls, New York, USA. 2009.

[9]   D. Park; J. Park, "An Enhanced Ranging Scheme Using WiFi RSSI Measurements for Ubiquitous Location," Computers, Networks, Systems and Industrial Engineering (CNSI), 2011 First ACIS/JNU International Conference on , vol., no., pp.296,301, 23-25 May 2011

[10]   Huang, Norden E. "Applications of Hilbert–Huang transform to non-stationary financial time series analysis." Applied Stochastic Models in Business and Industry 19.3 (2003): 245-268.

[11]   Z. Wu and H. Norden E. "Ensemble empirical mode decomposition: a noise-assisted data analysis method." Advances in Adaptive Data Analysis 1.01 (2009): 1-41.

[12]   I. J. Q. Binghao Li, Andrew G. Dempster,, "On outdoor positioning with Wi-Fi," *Journal of Global Positioning System,* vol. 7, pp. 18-26, 2008.

[13]   S. Saha, K. Chaudhuri, D. Sanghi, and P. Bhagwat, "Location determination of a mobile device using IEEE 802.11b access point signals," in *Wireless Communications and Networking, 2003. WCNC 2003. 2003 IEEE,* 2003, pp. 1987-1992 vol.3.

[14]   X. C. Yuan Song, Yoo-Ah Kim, Bing Wang, Hieu Dinha, Guanling Chen, "Sniffer channel selection for monitoring wireless LANs," *Computer Communications,* vol. 35, pp. 1994–2003, 2012.

[15]   U. Deshpande, T. Henderson, and D. Kotz, "Channel Sampling Strategies for Monitoring Wireless Networks." in *Modeling and Optimization in Mobile, Ad Hoc and Wireless Networks, 2006 4th International Symposium on,* 2006, pp. 1-7.

Permissions

All chapters in this book were first published in IJWMN, by AIRCC Publishing Corporation; hereby published with permission under the Creative Commons Attribution License or equivalent. Every chapter published in this book has been scrutinized by our experts. Their significance has been extensively debated. The topics covered herein carry significant findings which will fuel the growth of the discipline. They may even be implemented as practical applications or may be referred to as a beginning point for another development.

The contributors of this book come from diverse backgrounds, making this book a truly international effort. This book will bring forth new frontiers with its revolutionizing research information and detailed analysis of the nascent developments around the world.

We would like to thank all the contributing authors for lending their expertise to make the book truly unique. They have played a crucial role in the development of this book. Without their invaluable contributions this book wouldn't have been possible. They have made vital efforts to compile up to date information on the varied aspects of this subject to make this book a valuable addition to the collection of many professionals and students.

This book was conceptualized with the vision of imparting up-to-date information and advanced data in this field. To ensure the same, a matchless editorial board was set up. Every individual on the board went through rigorous rounds of assessment to prove their worth. After which they invested a large part of their time researching and compiling the most relevant data for our readers.

The editorial board has been involved in producing this book since its inception. They have spent rigorous hours researching and exploring the diverse topics which have resulted in the successful publishing of this book. They have passed on their knowledge of decades through this book. To expedite this challenging task, the publisher supported the team at every step. A small team of assistant editors was also appointed to further simplify the editing procedure and attain best results for the readers.

Apart from the editorial board, the designing team has also invested a significant amount of their time in understanding the subject and creating the most relevant covers. They scrutinized every image to scout for the most suitable representation of the subject and create an appropriate cover for the book.

The publishing team has been an ardent support to the editorial, designing and production team. Their endless efforts to recruit the best for this project, has resulted in the accomplishment of this book. They are a veteran in the field of academics and their pool of knowledge is as vast as their experience in printing. Their expertise and guidance has proved useful at every step. Their uncompromising quality standards have made this book an exceptional effort. Their encouragement from time to time has been an inspiration for everyone.

The publisher and the editorial board hope that this book will prove to be a valuable piece of knowledge for researchers, students, practitioners and scholars across the globe.

List of Contributors

Huda Adibah Mohd Ramli
Faculty of Engineering and Information Technology, University of Technology, Sydney, Australia
International Islamic University Malaysia (IIUM), Kuala Lumpur, Malaysia

Kumbesan Sandrasegaran
Faculty of Engineering and Information Technology, University of Technology, Sydney, Australia

Riyaj Basukala
Faculty of Engineering and Information Technology, University of Technology, Sydney, Australia

Rachod Patachaianand
Faculty of Engineering and Information Technology, University of Technology, Sydney, Australia

Toyoba Sohana Afrin
Faculty of Engineering and Information Technology, University of Technology, Sydney, Australia

Md. Munjure Mowla
Department of Electrical & Electronic Engineering, Rajshahi University of Engineering & Technology, Rajshahi, Bangladesh

S.M. Mahmud Hasan
Department of Electronics & Telecommunication Engineering, Rajshahi University of Engineering & Technology, Rajshahi, Bangladesh

Kuo-Tsang Huang
Department of Electrical Engineering, Chang Gung University, Tao-Yuan, Taiwan

Yi-Nan Lin
Department of Electronic Engineering, Ming Chi University of Technology, New Taipei City, Taiwan

Jung-Hui Chiu
Department of Electrical Engineering, Chang Gung University, Tao-Yuan, Taiwan

Akhtar Husain
Department of Computer Science & Information Technology, MJP Rohilkhand University, Bareilly, India

Ram Shringar Raw
School of Computer and Systems Sciences, Jawaharlal Nehru University, New Delhi, India

Brajesh Kumar
Department of Computer Science & Information Technology, MJP Rohilkhand University, Bareilly, India

Amit Doegar
Department of Computer Science, NITTTR, Chandigarh, India

Hareesh.K
Research Scholar, Jawaharlal Nehru Technological University, Anantapur, A.P, India

Manjaiah D.H
Professor and Chairman, Department of CS, Mangalore University, Mangalore, India

Samir A. Elsagheer Mohamed
Electrical Engineering Department, Faculty of Engineering, South Valley University, Aswan, Egypt
Currentlly affiliated to the College of Computer, Qassim University, P.O.B 6688, Buryadah 51453, Qassim, KSA

A. Nasr
Radiation Engineering Dept., NCRRT, Atomic Energy Authority, Egypt
Currentlly affiliated to the College of Computer, Qassim University, P.O.B 6688, Buryadah 51453, Qassim, KSA

Gufran Ahmad Ansari
Currentlly affiliated to the College of Computer, Qassim University, P.O.B 6688, Buryadah 51453, Qassim, KSA

Murali Parameswaran
Department of Computer Science & Information Systems, BITS-Pilani, Pilani, India

Chittaranjan Hota
Department of Computer Science & Information Systems, BITS-Pilani, Hyderabad, India

Mr. Mayur C. Akewar
Department of Computer Science & Engineering, Shri Ramdeobaba College of Engineering and Management, Nagpur University, Nagpur, India

Dr. Nileshsingh V. Thakur
Department of Computer Science & Engineering, Shri Ramdeobaba College of Engineering and Management, Nagpur University, Nagpur, India

Vanaja Ramaswamy
Department Of Computer Science and Engineering, Sri Venkateswara College Of Engineering, Sriperumpudur taluk, Tamilnadu, India

Abinaya Sivarasu
Department Of Computer Science and Engineering, Sri Venkateswara College Of Engineering, Sriperumpudur taluk, Tamilnadu, India

Bharghavi Sridharan
Department Of Computer Science and Engineering, Sri Venkateswara College Of Engineering, Sriperumpudur taluk, Tamilnadu, India

Hamsalekha Venkates
Department Of Computer Science and Engineering, Sri Venkateswara College Of Engineering, Sriperumpudur taluk, Tamilnadu, India

Samir A. Elsagheer Mohamed
Electrical Engineering Department, Faculty of Engineering, South Valley University, Aswan, Egypt
In sabbatical leave in College of Computer, Qassim University, P.O.B 6688, Buryadah 51453, KSA

Dr. T S R Murthy
Professor,Shri Vishnu Engg. College for Women,Bhimavaram. West Godavari dist,A.P,India

D. Siva Rama Krishna
Asst. professor,Swarnandhra Institute of Engg. & Tech.,Narsapuram. West Godavari dist, A.P,India

Taner Arsan
Department of Computer Engineering, Kadir Has University, Istanbul, Turkey

Fatih Günay
Department of Computer Engineering, Kadir Has University, Istanbul, Turkey

Elif Kaya
Department of Computer Engineering, Kadir Has University, Istanbul, Turkey

Anis Charrada
CSE Research Unit, Tunisia Polytechnic School, Carthage University, Tunis, Tunisia

Abdelaziz Samet
CSE Research Unit, Tunisia Polytechnic School, Carthage University, Tunis, Tunisia

Vasil Hnatyshin
Department of Computer Science, Rowan University, Glassboro, NJ, USA

Zhuliang Xu
Faculty of Engineering and Information Technology University of Technology Sydney Sydney, Australia

K. Sandrasegaran
Faculty of Engineering and Information Technology University of Technology Sydney Sydney, Australia

Xiaoying Kong
Faculty of Engineering and Information Technology University of Technology Sydney Sydney, Australia

Xinning Zhu
Faculty of Engineering and Information Technology University of Technology Sydney Sydney, Australia

Jingbin Zhao
Faculty of Engineering and Information Technology University of Technology Sydney Sydney, Australia

Bin Hu
Faculty of Engineering and Information Technology University of Technology Sydney Sydney, Australia

Cheng-Chung Lin
Faculty of Engineering and Information Technology University of Technology Sydney Sydney, Australia

9 781682 851333